An outline of energy metabolism in man

An outline of energy metabolism in man

Gordon L Atkins

Department of Biochemistry, University of Edinburgh Medical School

William Heinemann Medical Books Limited · London

William Heinemann Medical Books Ltd
23 Bedford Square
London WC1B 3HH

First published 1981

Copyright © Gordon L Atkins 1981

ISBN 0 433 00950 0

Typeset by D. P. Media Limited, Hitchin, Herts.
Printed in Great Britain by
R. J. Acford Ltd, Chichester, West Sussex

Contents

Preface

1. Basic principles — 1
    1.1 General outline of metabolism — 3
    1.2 Major fuels and their contribution to energy production — 5
    1.3 High energy compounds — 7
    1.4 Uses of ATP — 9
    1.5 Cell structure and energy metabolism — 11

2. Catabolism of simple units — 13
    2.1 Overview of catabolism — 15
    2.2 Oxidative phosphorylation — 17
    2.3 The citric acid cycle — 19
    2.4 β-oxidation of free fatty acids — 21
    2.5 Glycolysis — 23
    2.6 Anaerobic glycolysis — 25
    2.7 Amino acid catabolism — 27

3. Resynthesis of simple units — 29
    3.1 Gluconeogenesis — 31
    3.2 Free fatty acid synthesis — 33
    3.3 Sources of NADPH for free fatty acid synthesis — 35
    3.4 Ketone body metabolism — 37
    3.5 Lipoprotein metabolism — 39

4. Storage forms — 41
    4.1 Triacylglycerols — 43
    4.2 Glycogen — 43
    4.3 Protein — 45

5. Control mechanisms — 47
    5.1 Control by hormones — 49
    5.2 Control by key enzymes — 51
    5.3 Control by enzyme concentration — 53

6. Integration of pathways within cells — 55
    6.1 Metabolism by different tissues — 57
    6.2 Oxidation of free fatty acids — 59
    6.3 Formation of ketone bodies — 61
    6.4 Oxidation of ketone bodies — 63
    6.5 Oxidation of amino acids — 65
    6.6 Gluconeogenesis from amino acids — 67
    6.7 Oxidation of glucose — 69
    6.8 Conversion of glucose into free fatty acids — 71
    6.9 Glycogen metabolism — 73

7. Whole body metabolism   75
    7.1 Whole body metabolism   77
    7.2 The fasted state: 6 to 24 hours   79
    7.3 The fasted state: 2 to 4 days   81
    7.4 The fasted state: over 2 weeks   83
    7.5 Fasted and exercised   85
    7.6 The fed state   87

Appendix   89
    A.1 Some essential compounds for energy metabolism   91

Some suggestions for further reading   92

Index   93

Preface

This account of energy metabolism is based on a set of lectures that I have given for several years as part of a course organised by the Postgraduate Board of Medicine, University of Edinburgh for the benefit of anaesthetists preparing for Part 1 of the FFA RCS. I have been prompted to write it partly because many of those attending the course have suggested that a book such as this would be useful to them and none is currently available. I am also aware that energy metabolism is taught at this level to many students who are not biochemists, for example anaesthetists, nutritionists, nurses, etc. This book is also written for them.

In the presentation of the material, the basic principle I wish to convey is that material flows smoothly, and in a controlled manner, through several pathways to achieve certain objectives. I wish to avoid rigid compartmentalization of the subject, which most textbooks employ, and to do this I make extensive use of diagrams. As a result, this account is only a framework or outline with most of the detail, especially enzymes and less important intermediates, omitted.

The presentation here follows the pattern of the lecture series. It is a logical development in which, first, the various conventional pathways available to a cell for the metabolism of the three major fuels in man (carbohydrate, fat and protein) are discussed. Then there is a short account of the more important control mechanisms and this is followed by a discussion of the overall pathways in relation to the specialized functions of the various organs. Finally, the objective of the book is reached and the metabolism of the body as a whole is presented.

Author's note

Since this book was written the compound *fructose diphosphate* has become known as *fructose bisphosphate*.

1. BASIC PRINCIPLES

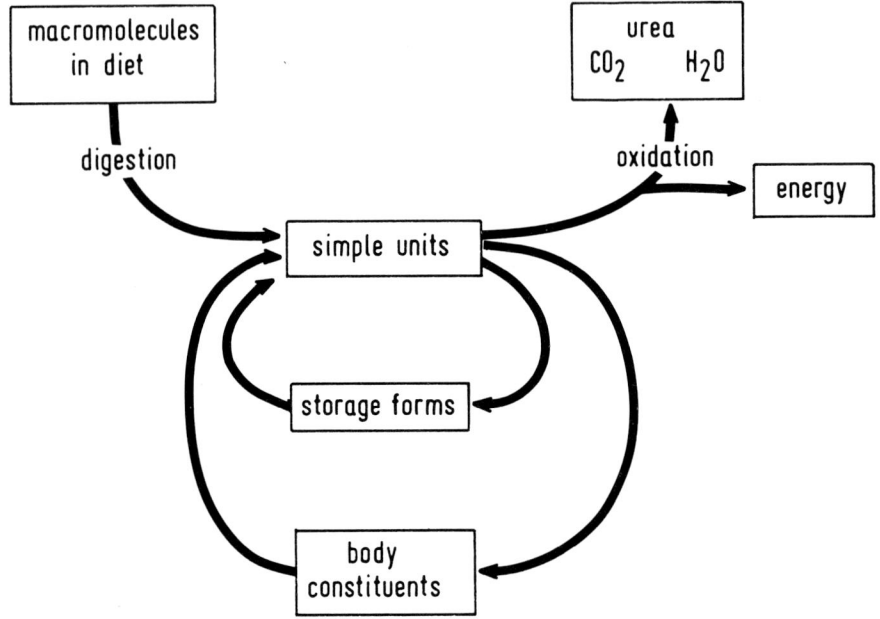

Fig. 1.1 *General outline of metabolism*

1.1 General outline of metabolism

The body, in order to survive, must have a continuous supply of energy. This energy is supplied by the internal combustion, i.e. oxidation, of exogenous organic molecules. In addition to an adequate supply of fuel sufficient oxygen must be available. The metabolic fuels are supplied as macromolecules in the diet, with fat, carbohydrates and protein supplying most of the energy. Internal metabolism and transport are, however, mainly as relatively simple units.

Digestion is the hydrolysis of macromolecules after which they are absorbed as various simple units, although fats are resynthesized immediately after absorption. Digestion serves two purposes. First, largely insoluble compounds are converted into soluble ones that can be easily transported in the body and absorbed by cells. Secondly, a very large number of different types of macromolecule (perhaps about 1000) are reduced to a relatively small number (about 50) of simple units. Digestion is otherwise of little importance, energetically, because only small amounts of energy are involved and these are lost as heat.

Within the body the various cells can generally only absorb or secrete simple units, although there are exceptions e.g. the lipoproteins. The further metabolism of these simple units depends on the balance between fuel supplied and energy needed. The possible routes are:

(i) oxidation to provide energy with N excreted as urea, C as carbon dioxide and H as water;
(ii) conversion of excess into various storage forms;
(iii) synthesis of internal cell structures, e.g. membranes, enzymes, etc.

Simple units are also obtained by internal catabolism. Catabolism is the conversion of a complex molecule into simpler ones; the term will therefore include oxidation. There is a constant turnover of cell constituents, and also storage compounds are hydrolysed when there is a fuel deficit.

In summary, the source of the metabolic fuels is either from intestinal absorption during the digestion of a meal (absorptive phase) or from the breakdown of internal stores (special stores or cell constituents) during other times.

Type of fuel	Forms in food	Simple units	Storage forms	Approximate daily intake	kJ/g	kJ/day	%
Triacylglycerols	triacylglycerols	free fatty acids acetoacetate β-hydroxybutyrate	triacylglycerols	160 g	39	6100	61
Carbohydrate	starch sucrose lactose	glucose fructose galactose lactate	glycogen	160 g	17	2700	27
Protein	protein	amino acids	protein	70 g	17	1200	12
					Totals	10 000	100

Table 1.2 *Major fuels and their contribution to energy production*

1.2 Major fuels and their contribution to energy production

Table 1.2 shows, for the three major types of fuel, their food forms, their storage forms and the simple units into which they are hydrolysed. The energy requirements and the proportions of the three types will vary according to sex, age and occupation. The daily intake rate shown is approximately that to satisfy a young sedentary male.

Notice that fat supplies about two-thirds of the energy needs whereas protein is relatively unimportant for that purpose.

Compound	Abbreviation
adenosine triphosphate	ATP
phosphocreatine	PC
nicotinamide adenine dinucleotide	NAD⁺ (oxidized)
	NADH (reduced)
nicotinamide adenine dinucleotide phosphate	NADP⁺ (oxidized)
	NADPH (reduced)
flavoprotein: FAD-protein	fp (oxidized)
(FAD = flavine adenine dinucleotide)	fpH_2 (reduced)

Table 1.3 *High energy compounds*

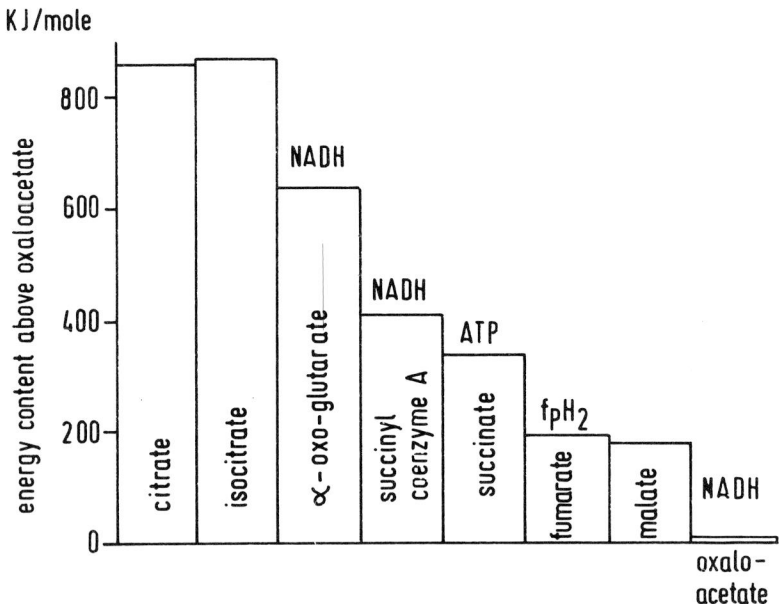

Fig. 1.3 *Energy conservation in the citric acid cycle*

1.3 High energy compounds

ATP is the most important high energy compound in the body. The majority of conserved energy is channelled into this compound so that most reactions or processes that need energy will, therefore, use ATP. The energy is stored as a high energy bond by the process

$$\text{energy} + \text{phosphate} + \text{ADP} \rightarrow \text{ATP}$$

(ADP = adenosine diphosphate). Energy is provided by the reverse of the reaction.

Phosphocreatine is used in several tissues (muscle, brain) as a temporary or back-up store for the energy conserved in ATP by transfer of the high energy phosphate group to creatine:

$$\text{creatine} + \text{ATP} \rightarrow \text{phosphocreatine} + \text{ADP}$$

Again the energy can be recovered by the reverse of the reaction.

NADH, NADPH and fpH_2 are intermediate forms in the conservation of energy as ATP. These three compounds are examples of coenzymes, i.e. they transfer small chemical groups (in this case hydrogen atoms) from one substrate or pathway to another. These coenzymes are normally reoxidized to generate ATP, but sometimes (especially in synthesis) they are used directly. For example, this occurs in free fatty acid synthesis from glucose (section 3.3). The equivalents in terms of ATP are:

$$1 \text{ NADH} = 1 \text{ NADPH} = 3 \text{ ATP}$$
$$1 \text{ } fpH_2 = 2 \text{ ATP}$$

During a sequence of catabolic reactions the energy content of the various reactants usually falls. If a large enough drop occurs during a reaction energy can be saved, and the type of high energy compound involved in the conservation depends on the size of the drop. For a large drop NADH is formed and for a small one the product is ATP. For an intermediate drop the high energy compound is fpH_2. If the drop is too small then energy is lost as heat. Fig. 1.3 shows the energy contents of the intermediates of the citric acid cycle (section 2.3). There are three steps involving NADH, one with fpH_2 and one with ATP.

The efficiency of energy conservation is therefore very high. In metabolism generally it is about 60%, whereas for man-made devices (petrol engines, etc.) it is very much lower. The prime reason is that catabolism involves a number of small and simple reaction steps and the type of high energy compound involved has been matched to the size of the energy drop of each reaction.

Energy conversion	Process
mechanical work	muscular contraction
active transport of compounds against a concentration gradient	ATPases in membranes e.g. Na^+/K^+ pumps
nervous activity	synthesis of transmitter compounds e.g. acetyl choline
chemical work	synthesis of storage compounds e.g. glycogen synthesis of cell constituents e.g. enzymes

Table 1.4 *Uses of ATP*

1.4 Uses of ATP

Table 1.4 summarizes the major types of process that use ATP directly. The relative amounts used depend on the activity of the individual. In resting man about one-third of the total energy used is by the brain and nervous tissue for ion transport and the synthesis of transmitter compounds.

The turnover of ATP is very rapid. Approximately 100 mole of ATP may be used each day, but the amount in the body at any given moment is only about 25 millimole. ATP, if the sole source of energy, would therefore last only about 30 seconds.

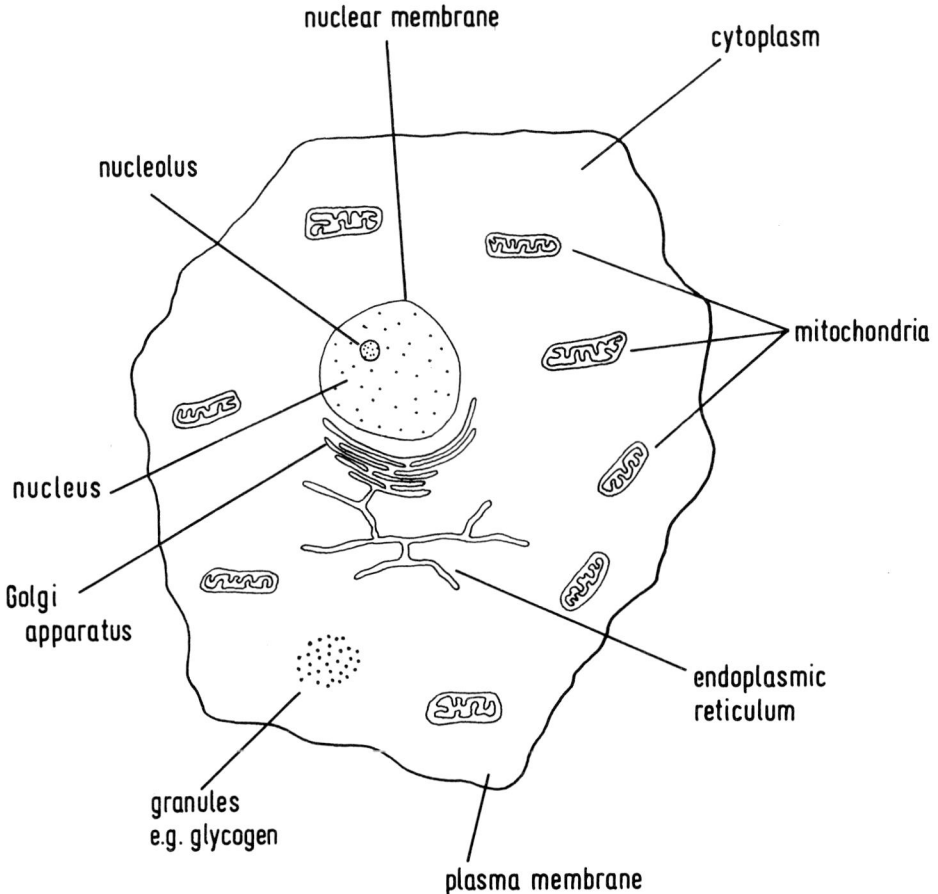

Fig. 1.5 *The more common structures within a typical cell*

1.5 Cell structure and energy metabolism

The basic structural unit in all tissues is the cell. The various organs differ from each other because their cells have a unique structure and a specialized biochemistry. These, in turn, determine the properties of the tissue. In spite of these differences, however, all the cells have certain features in common. These are indicated in the typical cell of Fig. 1.5.

The *plasma membrane* contains the cell and its contents. It has different permeabilities to different compounds. Thus some compounds are able to diffuse freely through the membrane, examples being glycerol and free fatty acids. Others, such as glucose, need a carrier mechanism; and in some tissues this type of transport can be controlled by other agents, for example glucose transport into muscle cells is determined by the concentration of insulin. Other compounds, such as the various coenzymes of energy metabolism, cannot diffuse out and are thus retained by the cell. The plasma membrane therefore determines to a large extent what materials associated with energy metabolism can enter and leave the cell and at what rates.

Within the cell the two other regions closely concerned with energy metabolism are the cytoplasm and the mitochondria. The *cytoplasm* contains water soluble enzymes and substrates and, as a generalization, is the area in which compounds like glucose, glycogen and free fatty acids are synthesized. The cytoplasm often also contains large regions of a membranous structure, the *endoplasmic reticulum*, which may be tubular in form. Some synthesis also occurs here, notably of phospholipids and proteins. The endoplasmic reticulum may be more compact in some places giving rise to the *Golgi apparatus*. The function of this structure seems to be to collect and perhaps process materials such as insulin and digestive enzymes before secretion.

The mitochondria are present in all cells (except the erythrocytes). The organelle has a double membrane, the inner one being folded inwards to present a large surface area. These infoldings are called *cristae*, and generally their number within the mitochondria corresponds approximately to the energy requirement of the cell. Thus mitochondria of red muscle tissue have many whereas those of white muscle have fewer. The organelle contains the enzymes of the citric acid cycle in the interior and the components of oxidative phosphorylation in the cristae. Therefore the major function of the mitochondria is to completely oxidize compounds and produce ATP. The membrane also exhibits different permeabilities, for example citrate and pyruvate can cross the membrane easily, oxaloacetate with difficulty and acetyl coenzyme A not at all. The membrane permeability thus has a profound effect on mitochondrial metabolism.

Other cell structures are less concerned with energy metabolism. For example, the nucleus and nucleolus contain the genetic information and the mechanism for controlling protein and enzyme synthesis.

2. CATABOLISM OF SIMPLE UNITS

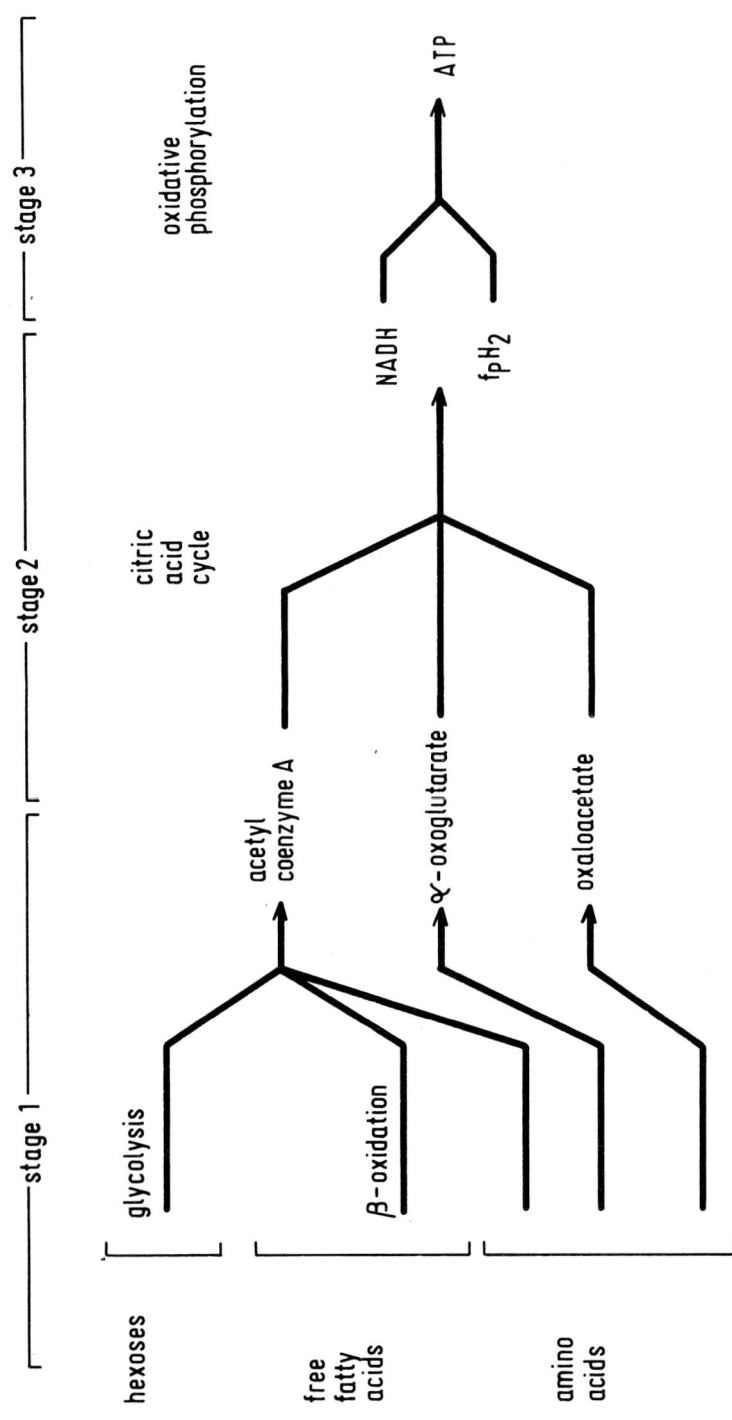

Fig. 2.1 *An overall view of catabolism*

2.1 Overview of catabolism

Catabolism of the simple units can be summarized, as in Fig. 2.1. The pathways of catabolism can be divided into three stages. Starting with a large number of compounds, each stage finishes with fewer. Thus progression through the three stages means that more and more material is channelled through fewer pathways until the last stage consists of only one pathway with two inputs.

Stage 1 takes a very large number of simple units and partially oxidizes them to give three major compounds, acetyl coenzyme A, α-oxoglutarate and oxaloacetate, and three minor ones, pyruvate, fumarate and succinyl coenzyme A. Many of the reaction sequences have sections in common – for example all the hexoses are fed into the earlier steps of glycolysis. Because this stage involves only partial oxidations, the energy released is only about one-third of the total possible.

Stage 2 is the complete oxidation of the six compounds from stage 1 to give carbon dioxide. This is the major route in the cell for energy trapping and the remaining two-thirds of the energy is released. It is conserved as NADH, fpH$_2$ and a little as ATP directly.

Stage 3 involves the reoxidation of the reduced coenzymes so that their hydrogens are released as water. The energy is finally transferred to ATP by the phosphorylation of ADP.

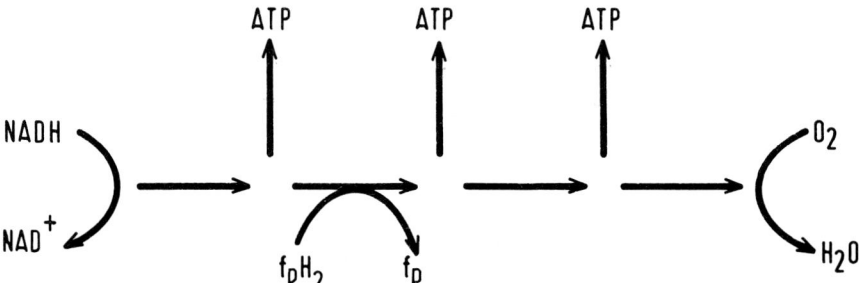

Fig. 2.2a *Oxidative phosphorylation*

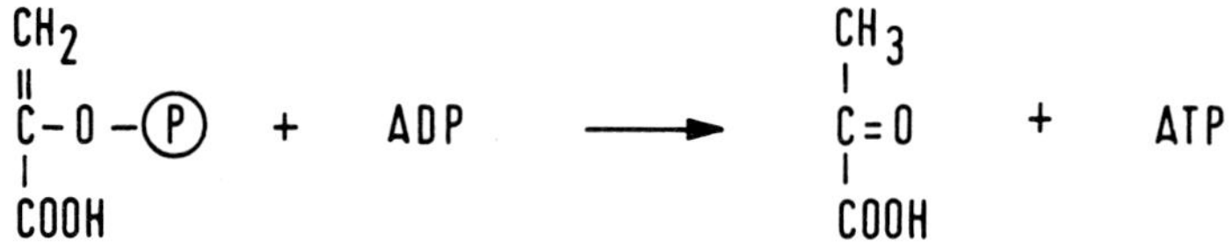

$$\begin{array}{c} CH_2 \\ \parallel \\ C-O-\text{\textcircled{P}} \\ \mid \\ COOH \end{array} \;+\; ADP \;\longrightarrow\; \begin{array}{c} CH_3 \\ \mid \\ C=O \\ \mid \\ COOH \end{array} \;+\; ATP$$

Fig. 2.2b *An example of substrate level phosphorylation*

2.2 Oxidative phosphorylation

This is stage 3 of section 2.1 and is outlined in Fig. 2.2a. The sequence of reactions is within the mitochondria and its function is the reoxidation of the reduced coenzymes, NADH to NAD^+ and fpH_2 to fp. Oxygen is involved at the last reaction and by receiving the hydrogens it is reduced to H_2O. During the sequence there are three places where the energy drop is large enough for the phosphorylation of ADP to ATP, i.e. for energy conservation (cf. section 1.3). Reoxidation of NADH causes three ATP to be formed. fpH_2 interacts at a later stage and one phosphorylation site is bypassed so that the reoxidation of fpH_2 yields only two ATP. Oxidative phosphorylation is responsible for about 95% of the ATP produced in the body and is the major site for oxygen utilization (about 90%).

The other 5% ATP is produced outside oxidative phosphorylation during reactions in which a compound has sufficient energy to cause the phosphorylation of ADP directly and no oxygen is involved. These reactions are called *substrate level phosphorylations*. The three main sites are phosphoenolpyruvate to pyruvate (Fig. 2.2b) and bisphosphoglycerate to 3-phosphoglycerate, both in glycolysis, and succinyl coenzyme A to succinate in the citric acid cycle. The first two reactions are important in that energy can be obtained by anaerobic glycolysis (section 2.6), i.e. when there is insufficient oxygen to meet an energy demand and oxidative phosphorylation cannot function at a fast enough rate.

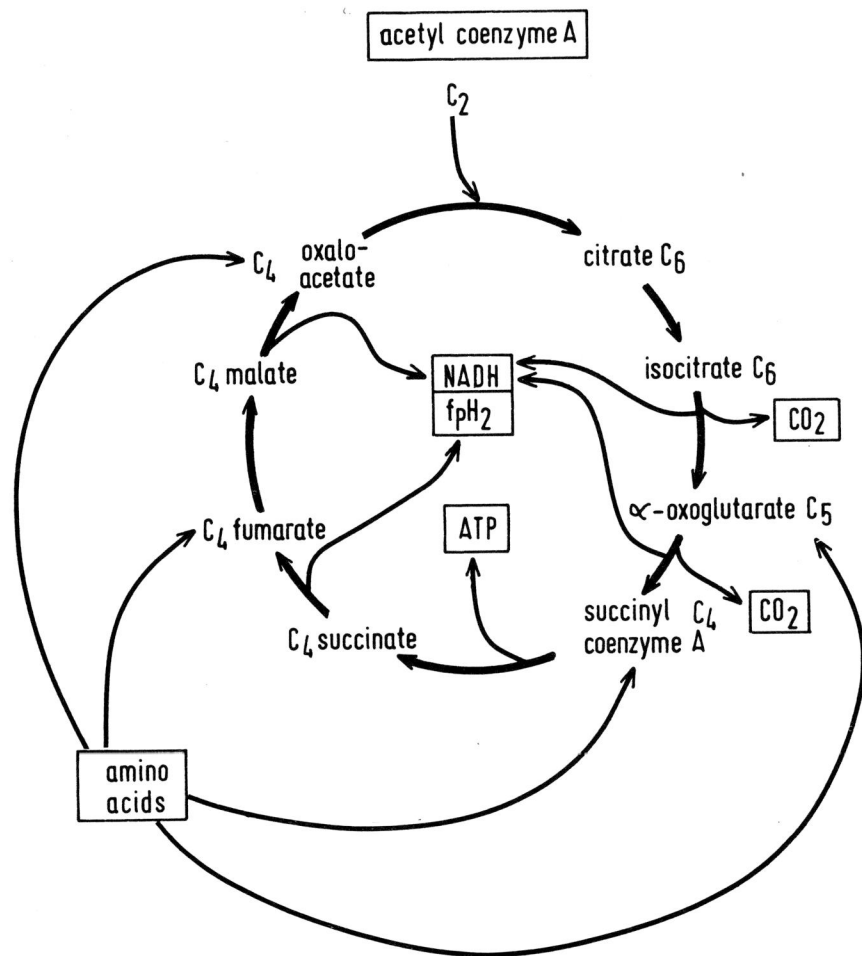

Fig. 2.3 *The citric acid cycle*

2.3 The citric acid cycle (Krebs cycle)

(Tricarboxylic acid cycle)

Stage 2 of section 2.1 is the citric acid cycle. Its major function is the provision of reduced coenzymes for stage 3. It is closely linked to oxidative phosphorylation because both pathways are in the mitochondria. The overall reaction is the complete oxidation of the compounds which enter so that carbon atoms become carbon dioxide and hydrogen becomes NADH and fpH$_2$. It is the major reaction sequence for the conservation of energy and is often called the terminal pathway for metabolism because the catabolism of all the simple units is channelled through this pathway.

The major input is acetyl coenzyme A from free fatty acids and hexoses. The sequence for the oxidation of acetyl coenzyme A (C$_2$) is briefly as follows. One oxaloacetate (C$_4$) is 'borrowed' from the cycle and with the acetyl coenzyme A (C$_2$) forms citrate (C$_6$). Passage round the cycle results in the loss of two C$_1$ units as CO$_2$, to finish with and return the 'borrowed' C$_4$ unit. The overall reaction, therefore, is the combustion of one acetate unit to two carbon dioxide molecules.

Amino acids are catabolized by various pathways and about ten of them, including most of the quantitatively important ones, enter the cycle via α-oxoglutarate, succinyl coenzyme A, fumarate or oxaloacetate (section 2.7).

Most of the other amino acids and acetoacetate enter via acetyl coenzyme A.

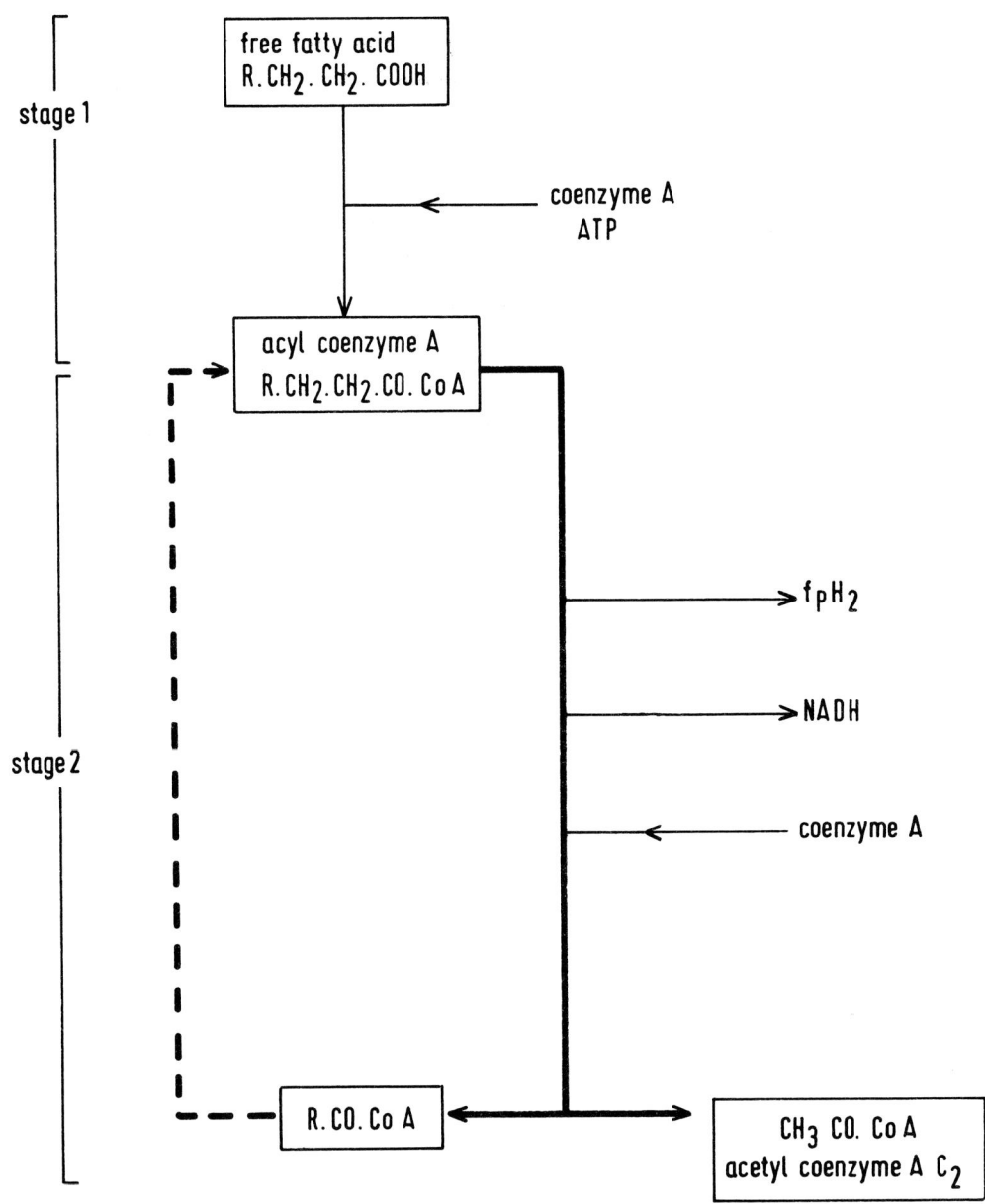

Fig. 2.4 *β-oxidation of free fatty acids*

2.4 β-oxidation of free fatty acids

This pathway is the major source of acetyl coenzyme A for oxidation by the citric acid cycle. It is in the mitochondria and thus proximate to the cycle. Most cells obtain their free fatty acids from the blood either through the free fatty acid-albumin complex or by the action of lipoprotein lipase on lipoproteins at the endothelial cell layer of the tissue.

The process can be divided into two stages. Stage 1 is an activation step whereby free fatty acid, with input of energy from ATP, is converted into acyl coenzyme A. Stage 2 is the oxidative cycle by which the acyl coenzyme A has two carbons removed as acetyl coenzyme A leaving another acyl coenzyme A but with two carbons less. This re-enters the cycle and because most natural free fatty acids have an even number of carbon atoms the final acyl coenzyme A is acetyl coenzyme A. The hydrogens are removed as NADH and fpH$_2$. The oxidation is only partial and the energy released in this pathway is much less than that obtained when the acetyl coenzyme A is finally and completely oxidized in the citric acid cycle.

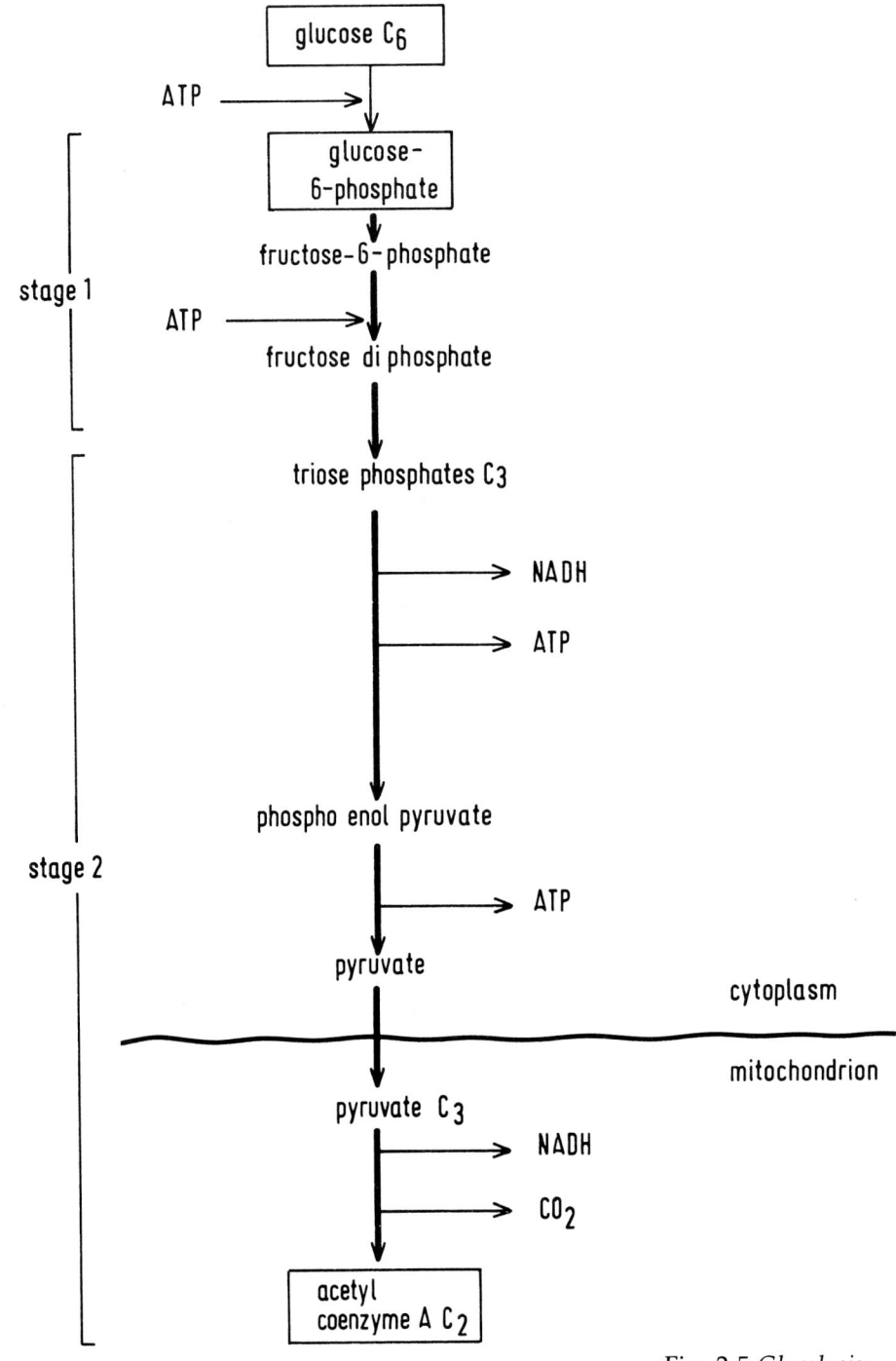

Fig. 2.5 *Glycolysis*

2.5 Glycolysis (the Embden-Meyerhof pathway)

Glycolysis is the second major source of acetyl coenzyme A and is the pathway by which hexoses are partially oxidized. The most important hexose is glucose and most cells obtain it from the blood. Its origin is either from absorption by the intestines during digestion or from gluconeogenesis or glycogenolysis by the liver during fasting.

Again there are two stages. The first, as far as fructose diphosphate, is activation with an input of energy from ATP. In the second stage the C_6 unit is split to give two C_3 units (the triose phosphates). These C_3 units are then partially oxidized. The major part of the sequence is in the cytoplasm, but the last step is mitochondrial. Transport across the mitochondrial membrane is as pyruvate. A certain amount of energy is conserved as NADH and ATP, the latter by substrate level phosphorylation (section 2.2), but again the amount of energy is much less than that obtained by the final oxidation of acetyl coenzyme A in the citric acid cycle.

Galactose is converted into glucose and fructose is phosphorylated by its own specific reaction to enter via the triose phosphates. Some amino acids enter as pyruvate and glycerol enters as triose phosphate.

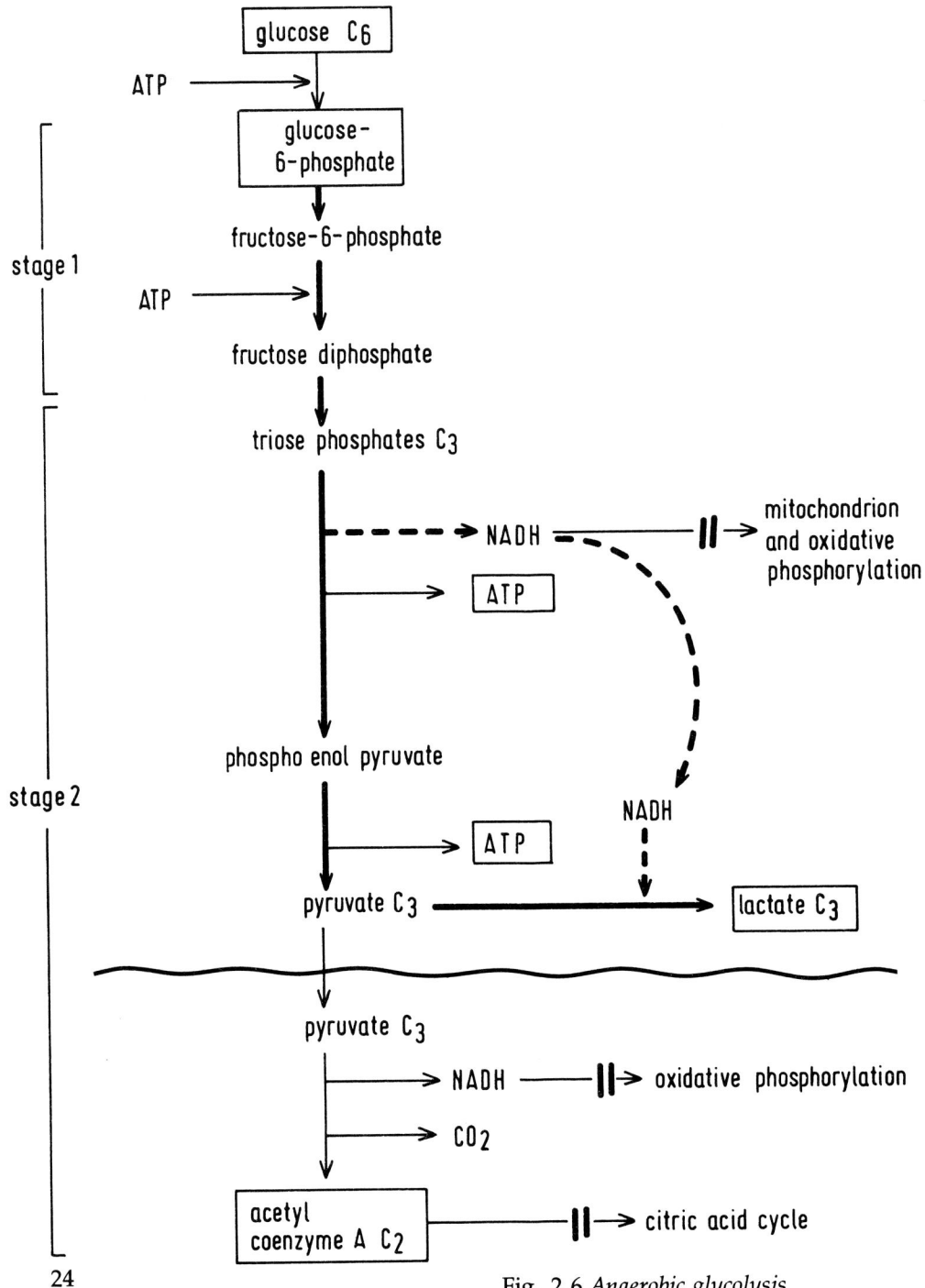

Fig. 2.6 *Anaerobic glycolysis*

2.6 Anaerobic glycolysis

If there is no oxygen then (i) NADH will not be reoxidized by oxidative phosphorylation and (ii) the rate through the citric acid cycle will stop. The result is that NADH would build up and the acetyl coenzyme A would not be removed and oxidized. Both problems are solved by using the NADH to convert pyruvate to lactate. The NADH is thus reoxidized so that glycolysis can continue, but with lactate as the end product rather than acetyl coenzyme A. ATP is produced by substrate level phosphorylation. Since free fatty acid catabolism and amino acid catabolism do not have this mechanism, glycolysis is the only reaction sequence able to operate anaerobically and produce energy. The lactate produced is secreted by the cells and then used by the liver to resynthesize glucose (section 3.1).

If a tissue is supplied with inadequate amounts of oxygen then energy will be obtained partly by anaerobic glycolysis and partly by aerobic catabolism.

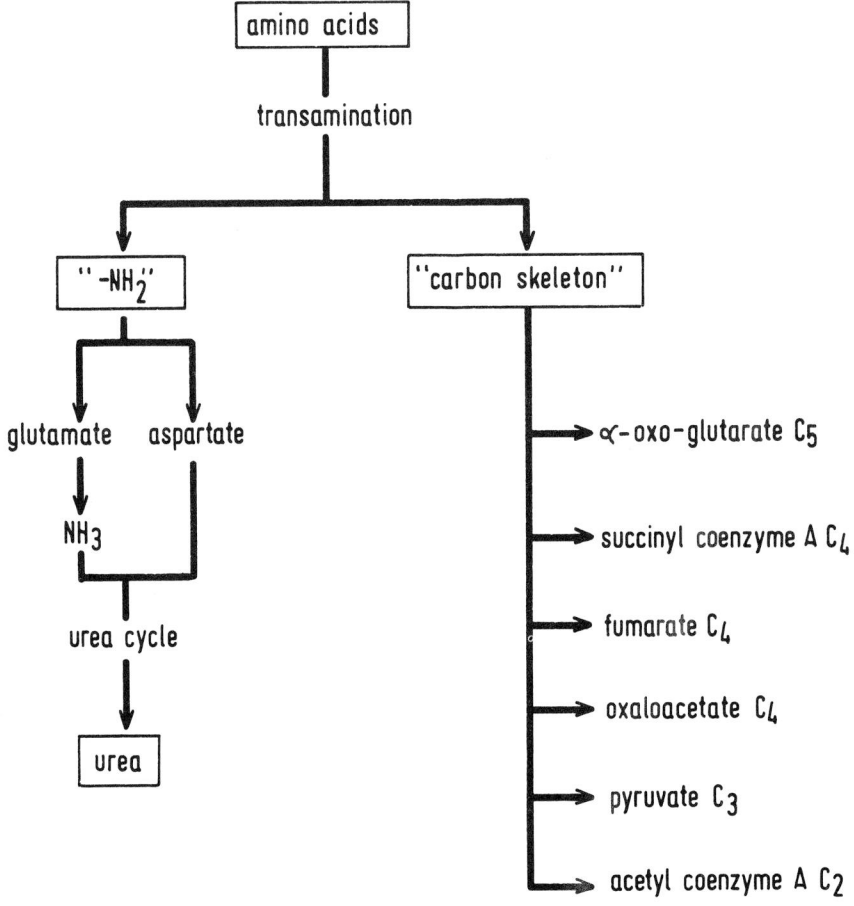

Fig. 2.7 *Amino acid catabolism*

2.7 Amino acid catabolism

For most amino acids the first step in catabolism is transamination, i.e. removal of the amino group:

amino acid + acceptor molecule → α-oxo acid + amino acid
(e.g. ('carbon (e.g.
α-oxoglutarate) skeleton') glutamate)

Amino groups can be passed though several amino acids but they all finish at glutamate and aspartate. Glutamate and aspartate are therefore intermediates in cyclic processes. Glutamate dehydrogenase produces ammonia which, together with aspartate and carbon dioxide, enters the urea cycle. Thus urea is formed. The synthesis of one molecule of urea requires much energy, four ATP, but on the other hand the end product is inert and water soluble so that it can be excreted easily.

Each amino acid has its own pathway for the catabolism of its carbon skeleton. These compounds are partially oxidized until eventually one of the intermediates of the citric acid cycle or glycolysis is reached.

3. RESYNTHESIS OF SIMPLE UNITS

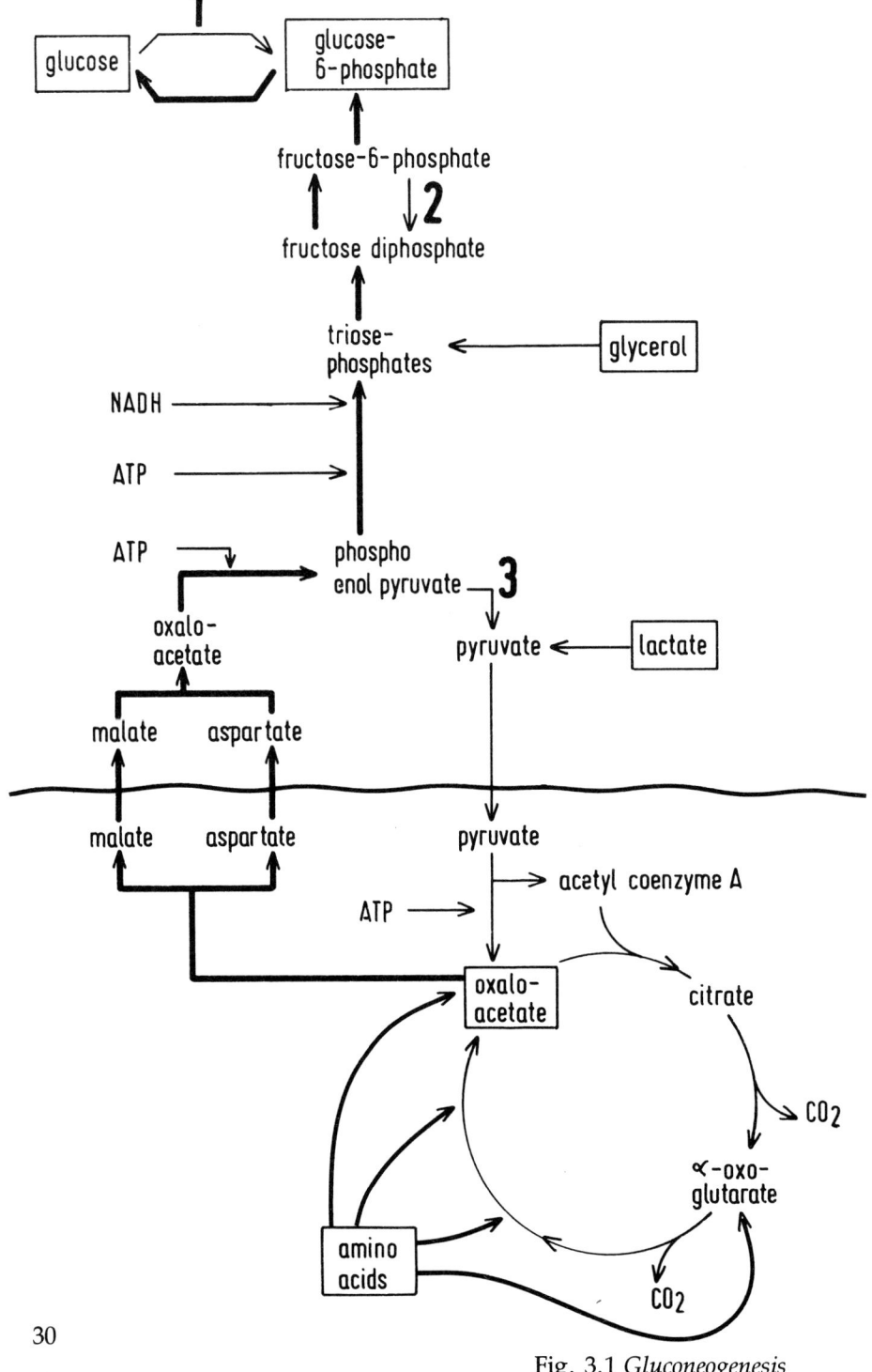

Fig. 3.1 *Gluconeogenesis*

3.1 Gluconeogenesis

Not all metabolism is concerned with the direct oxidation of simple units. Much of the flow of material involves converting one fuel into another. The latter fuel is often transported to another tissue where it may then be oxidized or stored. Processes are therefore present in cells for synthesizing the simple units from other intermediates in metabolism.

Gluconeogenesis is one such process and is defined as the resynthesis of glucose from smaller molecular weight precursors. The major site for this pathway is in the cytoplasm of liver, but it also occurs to a much smaller extent in kidney. Gluconeogenesis is important because it is one of the two major sources of glucose during fasting.

Most reactions in the pathway are the same as for glycolysis except for three steps. These are steps which would require too great an energy jump so that bypass reactions are necessary. Two steps (1 and 2) are bypassed by enzymes which hydrolyse phosphate links. The bypass for step 3 is more complex and involves the carboxylation of pyruvate to oxaloacetate, transport of oxaloacetate into the cytoplasm and then decarboxylation plus phosphorylation to give phosphoenolpyruvate. The transfer of oxaloacetate out of the mitochondria is via aspartate (after transamination) or malate (after hydrogenation), because oxaloacetate cannot diffuse rapidly enough.

The pathway is usually considered to start at oxaloacetate because precursors can be obtained from the citric acid cycle. The most important source of these is the amino acids. In liver the major amino acid is alanine (which enters via lactate) but in kidney it is glutamine (which enters via glutamate and α-oxoglutarate). Both these amino acids are released by skeletal muscle during fasting. The other two important precursors for gluconeogenesis are glycerol (from triacylglycerol hydrolysis) and lactate (from anaerobic glycolysis in other tissues). Lactate is not, however, a net source of glucose.

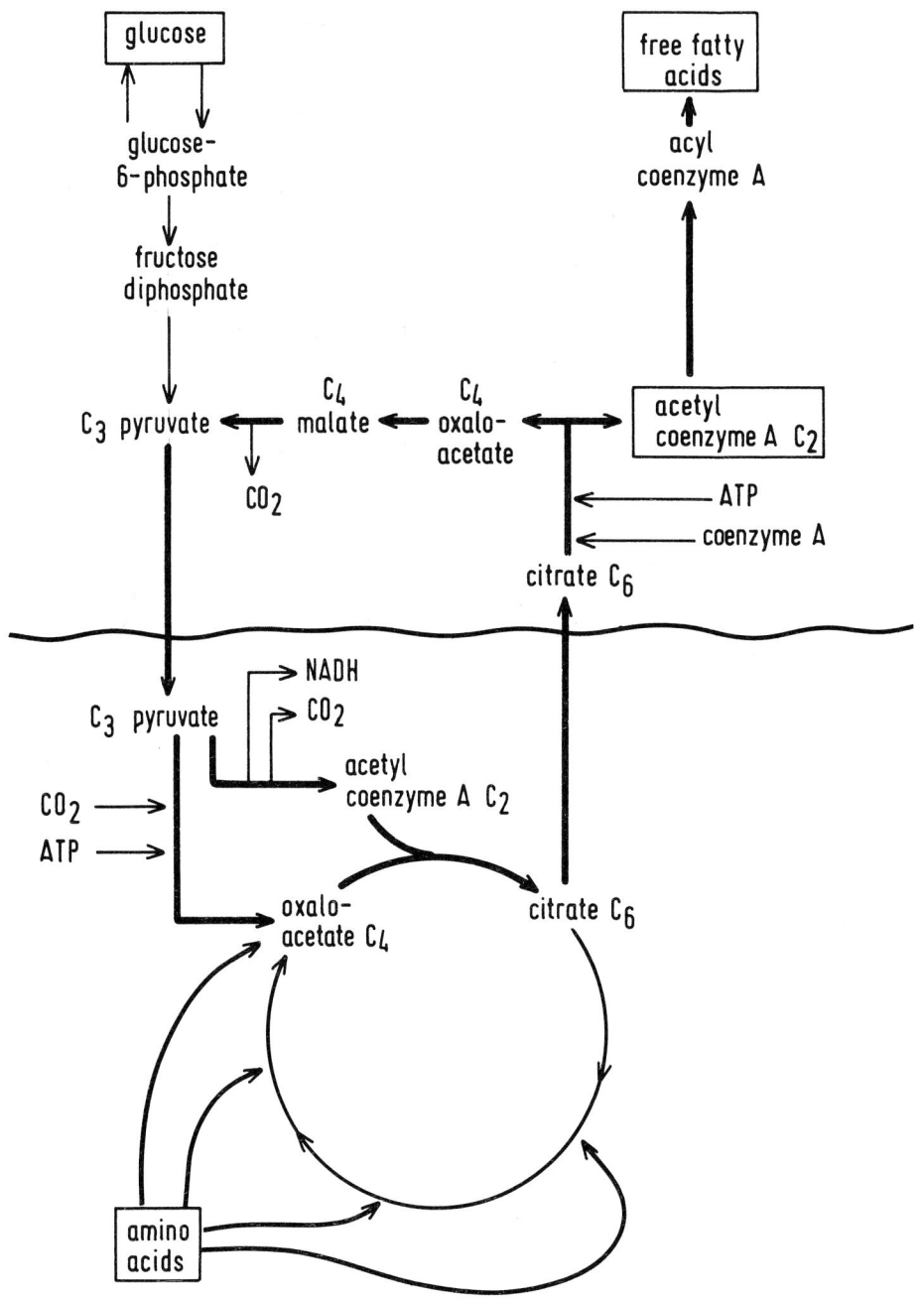

Fig. 3.2 *Free fatty acid synthesis*

3.2 Free fatty acid synthesis

This pathway is different to that for β-oxidation and occurs in the cytoplasm of cells. The starting point is acetyl coenzyme A and part of the sequence has similar types of reaction to β-oxidation except that NADPH provides the hydrogen for the reductive steps.

Acetyl coenzyme A is, however, needed in the cytoplasm whereas it is generated in the mitochondria. Transport out is by means of the citrate-malate cycle and the sequence is as follows. A pyruvate molecule enters the mitochondrion. A second pyruvate molecule is 'borrowed' from the cycle. The molecules ($2 \times C_3$) are then converted one each to acetyl coenzyme A (C_2). and oxaloacetate (C_4) followed by condensation to give citrate (C_6). Citrate diffuses out of the mitochondrion and is split, in the cytoplasm, into acetyl coenzyme A and oxaloacetate. Acetyl coenzyme A is thus available for fatty acid synthesis. The oxaloacetate is converted via malate into pyruvate which then returns to the mitochondrion to replace the 'borrowed' pyruvate. The overall reaction is

$$\text{pyruvate} \rightarrow \text{acetyl coenzyme A} + CO_2$$

This pathway is important because it is part of the process for storing surplus carbohydrates (glucose) and perhaps amino acids as triacylglycerols. Adipose tissue and liver are the two important organs for free fatty acid synthesis.

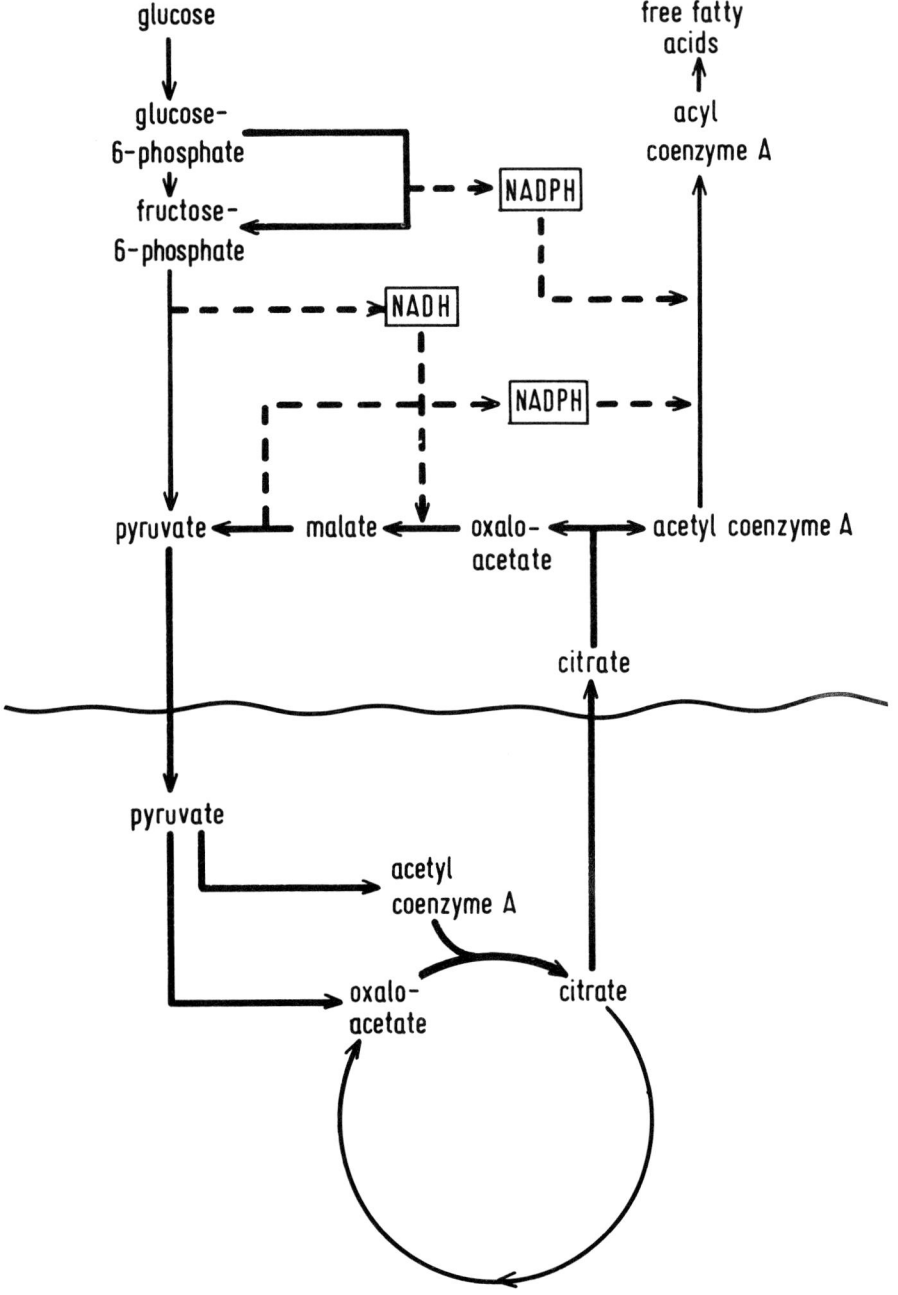

Fig. 3.3 *Sources of NADPH for free fatty acid synthesis*

3.3 Sources of NADPH for free fatty acid synthesis

Free fatty acid synthesis requires NADPH and there are two major sources.

The first is the citrate-malate cycle (section 3.2). Conversion of oxaloacetate to malate will reoxidize NADH (from glycolysis) to give NAD^+ for re-use. The next reaction, malate to pyruvate, reduces $NADP^+$ (from free fatty acid synthesis) and provides NADPH. The net result is that hydrogen is transferred from glycolysis to free fatty acid synthesis via the two coenzymes. One NADPH is formed for each acetyl coenzyme A transported out of the mitochondrion. This cycle therefore supplies half the NADPH required for free fatty acid synthesis.

The other half is supplied by the pentose phosphate pathway (hexose monophosphate shunt). The overall reaction is the complete oxidation of one C_6 unit and can be represented as

$$6 \text{ glucose-6-phosphate} \rightarrow 5 \text{ fructose-6-phosphate} + 6\ CO_2$$

This sequence, which is in fact a complex set of reactions, produces 12 NADPH for each C_6 unit oxidized.

The pentose phosphate pathway is cytoplasmic and its links with glycolysis are obvious. It is highly active in liver and adipose tissue, i.e. those tissues with a requirement for NADPH to support free fatty acid synthesis.

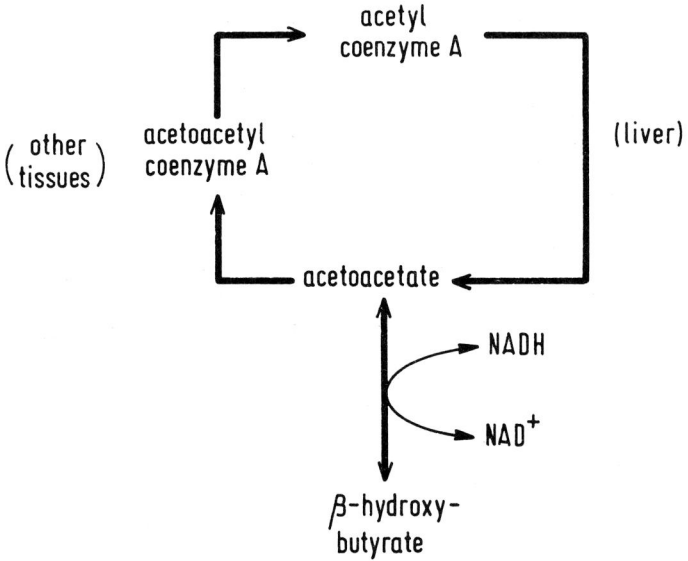

Fig. 3.4 *Ketone body metabolism*

3.4 Ketone body metabolism

The two important ketone bodies are acetoacetate and β-hydroxybutyrate. They are formed only in the liver and when large amounts of acetyl coenzyme A are present from β-oxidation. The reaction can be summarized (Fig. 3.4) as two acetyl coenzyme A units ($2 \times C_2$) condensing to give one acetoacetate (C_4). Reduction with NADH gives β-hydroxybutyrate. Liver is the only tissue which has the enzymes necessary for producing free acetoacetate and β-hydroxybutyrate.

The subsequent oxidation of these two compounds must occur in other tissues because only they, and not liver, can convert acetoacetate to acetoacetyl coenzyme A. The latter is then converted into acetyl coenzyme A which enters the citric acid cycle.

This is a mechanism whereby the liver converts one type of fuel into another type that can be used more easily by some other organ. For example, nervous tissue cannot use free fatty acids but can use ketone bodies during prolonged fasting.

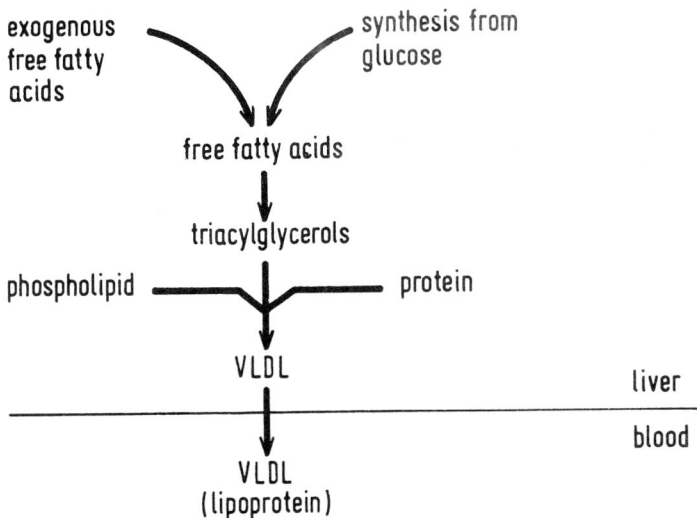

Fig. 3.5a *Synthesis and secretion of lipoprotein by the liver*

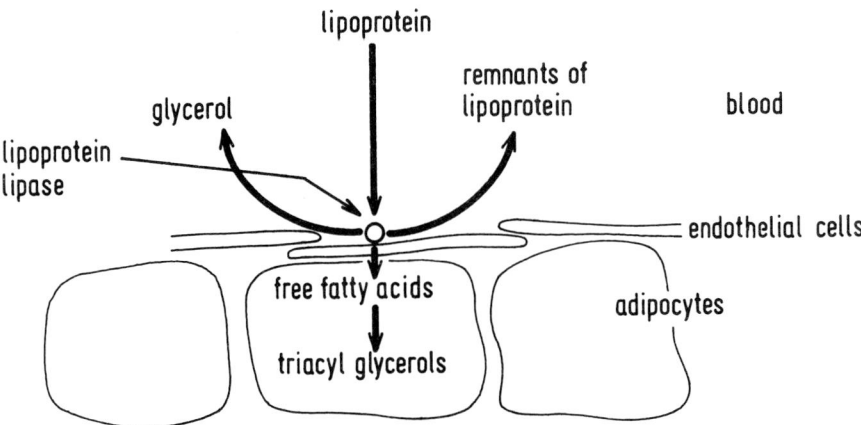

Fig. 3.5b *Uptake of free fatty acids from lipoprotein by adipose tissue*

3.5 Lipoprotein metabolism

Lipoproteins are not simple units, but because their major role is the transport of free fatty acids from the liver to adipose tissue it is convenient to consider them here.

Lipoproteins, or more precisely the group known as very low density lipoproteins (VLDL), are complexes of triacylglycerols, phospholipid, cholesterol and cholesterol esters with protein. Approximately 60% of the complex consists of triacylglycerols. Triacylglycerols are insoluble in water but complexing them with other lipids and protein produces soluble, stable dispersions which can be readily transported in blood.

The free fatty acids come from two sources, either synthesis from glucose within the liver or free fatty acids released by adipose tissue. The free fatty acids are synthesized into triacylglycerols which are then 'packaged' with the other components and secreted. Fig. 3.5a summarizes the process.

The three major tissues which take up lipoproteins are adipose tissue, and skeletal and cardiac muscle. At the endothelial surfaces of these tissues lipoprotein lipase hydrolyses the triacylglycerols within the complex. Free fatty acids are taken up by the cells and the other components, glycerol and the remains of the VLDL, are returned to the liver (Fig. 3.5b). In adipose tissue the lipoprotein lipase is sensitive to insulin and is active only during the absorption of food, so that the free fatty acids are channelled into storage as triacylglycerols. In muscle, the lipoprotein lipase is only active during fasting and the free fatty acids are channelled into oxidation.

4. STORAGE FORMS

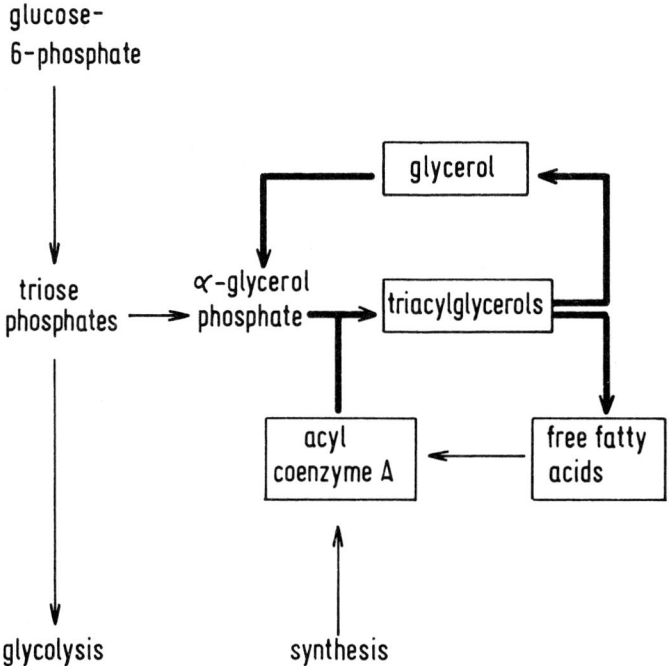

Fig. 4.1 *Triacylglycerol metabolism*

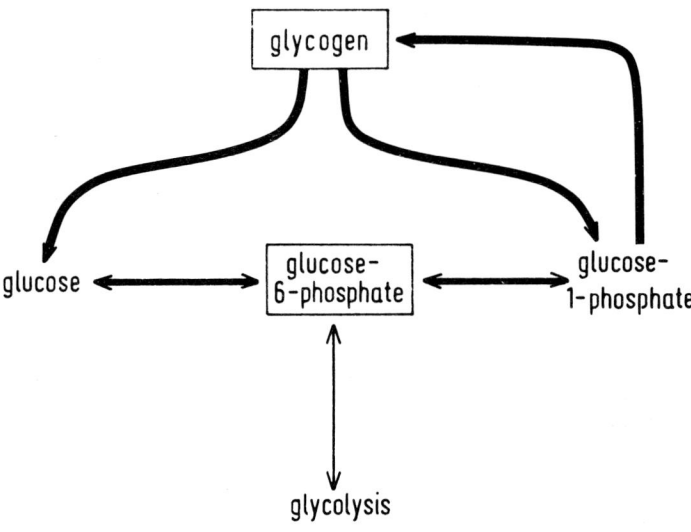

Fig. 4.2 *Glycogen metabolism*

4.1 Triacylglycerols

When the fuel supply is greater than the energy demand then the excess energy is stored. Sometimes one fuel is converted to another first. For example, glucose is often changed into free fatty acids then stored as triacylglycerols.

In fact, the major storage form for excess energy is the triacylglycerols. The reactions for synthesis are partly in the cytoplasm and partly in the endoplasmic reticulum whereas those for degradation are solely in the cytoplasm. Synthesis uses acyl coenzyme A and glycerol phosphate from glycolysis. Breakdown yields free fatty acids and free glycerol. As for most materials, the two pathways are therefore quite distinct.

The two important organs for the synthesis and degradation of triacylglycerols are adipose tissue and liver. Synthesis also occurs in the intestines during the absorption of fat. Liver and intestines can convert glycerol back to glycerol phosphate, but adipose tissue cannot, so that the glycerol is secreted with the free fatty acids.

4.2 Glycogen

Glucose can be stored directly as its polymer, glycogen. The pathways for the synthesis and degradation of glycogen are both cytoplasmic, but they use different enzymes. Synthesis proceeds via glucose-1-phosphate and uses the enzyme glycogen synthetase. Degradation, using phosphorylase, produces glucose-1-phosphate but other enzymes cause about 10% of glycogen to appear as free glucose. Again, two separate pathways are used for synthesis and degradation. It allows the two routes to be individually controlled (section 5.2).

Glycogen is synthesized when more glucose is available than required to produce energy and it is degraded when the body needs glucose.

Most tissues make and store glycogen, but there are two major ones, i.e. liver and muscle. Up to about 10% by weight of the liver can be glycogen. Muscle may have up to about 4% but, because of its greater mass, will store a greater weight. The glycogen store in the brain is relatively small and can support anaerobic glycolysis for only about four minutes.

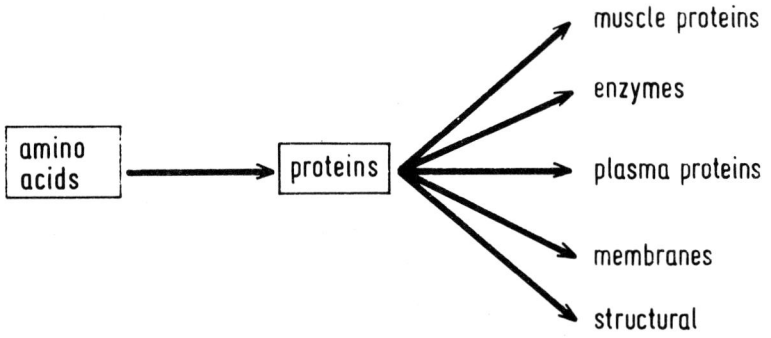

Fig. 4.3 *Synthesis of proteins*

4.3 Protein

A supply of amino acids will allow proteins to be synthesized. The process may be limited, however, by the availability of one or more of the essential amino acids (section A.1). Since essential amino acids cannot be synthesized a shortage of one of these will limit the rate of protein synthesis.

Whereas carbohydrate and fat have specific storage forms, glycogen and triacylglycerols, there is no special depot for protein. The proteins that are synthesized are those essential to the life of the cell or organ, i.e. muscle proteins, enzymes, serum proteins, connective tissue, etc.

Amino acids are needed for gluconeogenesis during fasting (section 3.1). Therefore it is the above proteins, especially those from visceral organs and muscle, that must be degraded to provide a source of the amino acids.

5. CONTROL MECHANISMS

Hormone	Release		Effect on		
	Stimuli	Inhibitors	Adipose	Muscle	Liver
Insulin	hyperglycaemia (amino acids) (FFA)	adrenaline	TAG uptake↑ glucose→TAG↑ FFA release↓	protein synthesis↑ glucose uptake↑ glycogen synthesis↑	glycogen synthesis↑ glucose release↓ KB synthesis↓
Glucagon	hypoglycaemia stress	FFA hyperglycaemia insulin			glycogenolysis↑ gluconeogenesis↑ AA catabolism↑ KB synthesis↑
Adrenaline	hypoglycaemia stress		FFA release↑ glucose uptake↑	FFA utilisation↑ glycogenolysis↑ glucose uptake↓	glycogenolysis↑ gluconeogenesis↑ KB synthesis↑
Growth hormone	hypoglycaemia stress	hyperglycaemia	FFA release↑ glucose uptake↑ FFA synthesis↓	protein synthesis↑ glucose uptake↓	glycogen synthesis↑ glucose release↑
Cortisol	hypoglycaemia injury/stress		FFA release↑	protein synthesis↓ glucose uptake↓ amino acid release↑	AA→glucose↑ glycogen synthesis↑ glucose release↑

Abbreviations:
AA   amino acids
FFA   free fatty acids
TAG   triacylglycerols
KB   ketone bodies: acetoacetate and β-hydroxybutyrate

Table 5.1 *The more important hormones controlling energy metabolism*

5.1 Control by hormones

The flow of the material through the various pathways of energy metabolism must be controlled so that any flow is in the desired direction and at the correct rate, e.g. so that excess glucose is converted into triacylglycerols and not oxidized. The three important methods for control are as follows. First, hormones control the flow of metabolic fuels between organs. Secondly, intermediates within the cell control the rates of various pathways. Thirdly, the maximum rate of a pathway is controlled by the amounts of enzymes present. These three controls act at different levels within the cell and over different time scales.

Table 5.1 lists the more important hormones that control energy metabolism together with the controls acting on their release. It also summarizes their major effects on the flow of materials between the three most important organs. Some of these are direct controls on transport processes or enzyme activities, some may be more indirect. Stress and injury also cause release of hormones and this is achieved by the action of the sympathetic nervous system.

The actions within the cells of some hormones are effected by an intermediary agent, cyclic AMP (3', 5' – adenosine monophosphate). As an example, adrenaline causes glycogenolysis in the liver by first binding at a membrane receptor site. This allows an enzyme, adenylate cyclase associated with the receptor site, to produce cyclic AMP from ATP. This in turn activates glycogen phosphorylase which leads to an increased rate for the breakdown of glycogen. The action of glucagon is identical to that of adrenaline.

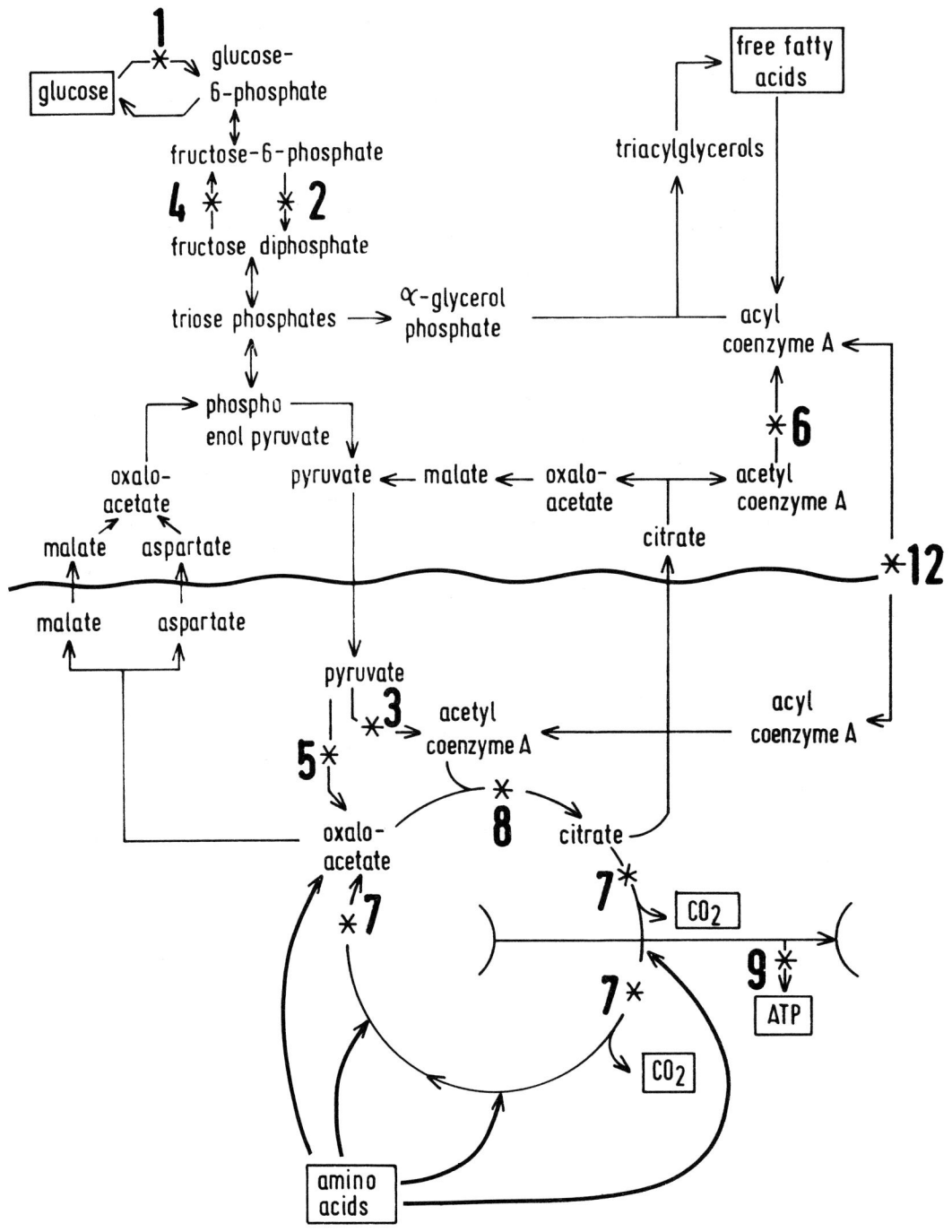

Fig. 5.2 *Control of the pathways for energy metabolism*

5.2 Control by key enzymes

Control of a pathway by key enzymes is by means of the concentrations of intermediates. Key enzymes are so called because they are usually found to be rate limiting under normal conditions of metabolism. An enzyme's activity is controlled by (i) displacing a near-equilibrium reaction from one position to another with a higher or lower rate; (ii) altering the enzyme's configuration so that its activity is greater or smaller; or (iii) by covalently modifying the enzyme's structure and thereby changing its activity. In this brief account distinctions are not made between the three modes of action.

Generally, overproduction of an end product will reduce the activity of the first enzyme in a sequence so that the production rate is reduced. Overproduction of an earlier metabolite may activate a later enzyme so that the metabolite is removed.

The major control for energy metabolism is by means of the energy demand within the cell. During energy production, material will be oxidized by the citric acid cycle and oxidative phosphorylation so that ATP is produced. Overproduction, i.e. a high ATP concentration, will inhibit oxidative phosphorylation at step 9 (Fig. 5.2) resulting in a build up of NADH. The increased concentration of NADH will in turn cause inhibition of the citric acid cycle at the various dehydrogenases that produce NADH (steps 7). High concentrations of ADP and phosphate will occur when there is a shortage of ATP and these will therefore stimulate oxidative phosphorylation.

AMP is also an important controlling agent because of the reaction

$$2\ ADP \rightleftharpoons AMP + ATP$$

During the rapid use of ATP the low concentration of ATP and the high concentration of ADP drive the reaction to the right hand side so that AMP is produced.

Table 5.2 (overleaf) summarizes some of the more important controls on key enzymes.

Pathway	Enzyme controlled		Inhibitors	Activators
glycolysis	hexokinase	1	glucose-6-phosphate	
	phosphofructokinase	2	ATP, citrate	ADP, AMP
	pyruvate dehydrogenase	3	acetyl coenzyme A	
gluconeogenesis	fructose diphosphatase	4	AMP	
	pyruvate carboxylase	5		acetyl coenzyme A
free fatty acid synthesis	acetyl coenzyme A carboxylase	6	acyl coenzyme A	citrate
citric acid cycle	dehydrogenases	7	NADH	AMP
	citrate synthetase	8	NADH	
oxidative phosphorylation	phosphorylation	9	ATP	ADP, phosphate
transfer of acyl units into mitochondria	acyl carnitine transferase	12		

Table 5.2 *Control of key enzymes*

5.3 Control of enzyme concentration

This control acts on the mechanism for synthesizing enzymes. An increased rate of synthesis will cause the concentration of an enzyme to increase also. A decreased rate of synthesis allows the concentration to fall. The control is relatively slow, needing a few days to take maximum effect, and is therefore important only for long term alterations of metabolism. For example, during fasting the liver enzymes for gluconeogenesis and the urea cycle will increase in amount and therefore increase the maximum rates possible for these pathways.

The stimulus is usually external to the cell. In the above example cortisol controls the rate of synthesis of these enzymes.

6. INTEGRATION OF PATHWAYS WITHIN CELLS

	Liver	Skeletal muscle	Cardiac muscle	Adipose	Brain CNS
Uptake, for energy or storage:					
lipoprotein	++	++	++	++	−
free fatty acids	++	++	++	−	−
glucose	++	++	+	++	++
lactate	++	−	++	−	−
amino acids	++	−	−	−	(+)
ketone bodies	−	++	++	−	(++)
Produced, used elsewhere:					
lipoprotein	++	−	−	−	−
free fatty acids	−	−	−	++	−
glucose	++	−	−	−	−
lactate	−	++	−	−	−
ketone bodies	++	−	−	−	−
Storage:					
triacylglycerols	+	+	+	++	−
glycogen	++	++	+	−	+

Table 6.1 *Summary of metabolism by different tissues*

6.1 Metabolism by different tissues

The pathways described so far indicate all the possibilities within a general type of cell. Liver cells are probably the only ones which approach this situation. The reason is that all cells are specialized and contain only the pathways necessary for their particular functions. In addition, not all the pathways are fully active at the same time. For example, synthesis and degradation of a given compound do not often occur simultaneously within the same tissue. If they do then the rates are usually quite different.

Table 6.1 summarizes the major differences between the important tissues involved in energy metabolism. Some additional comments follow.

Liver is a general purpose organ. It will take up most compounds and can perform most of the conversions between fuels.

The most important function of skeletal muscle is to oxidize compounds and produce mechanical work. It generally uses free fatty acids and ketone bodies in proportion to their particular concentrations in blood. Glucose is an important fuel only during hyperglycaemia and when the tissue becomes anaerobic. In this latter situation only glucose can be used and the lactate formed must be released because muscle lacks certain enzymes for gluconeogenesis. It is also the major source of amino acids, from protein, during fasting and releases these chiefly as alanine and glutamine. Storage of energy also occurs, mainly as glycogen and with smaller amounts as triacylglycerols.

Cardiac muscle is similar to skeletal muscle except that it is not normally called upon to work anaerobically. Lactate can, in fact, act as an energy source if it is present in blood. The main fuels are, therefore, free fatty acids, lactate and ketone bodies; and they are used in preference to glucose.

The chief function of adipose tissue is to store energy as fat. It converts glucose to triacylglycerols and also accepts external triacylglycerols for storage. It releases energy as free fatty acids and free glycerol.

Brain and the central nervous system normally use glucose as the sole source of energy. During fasting acetoacetate and β-hydroxybutyrate are important, however, and after three or four days they become the major fuels. Amino acids may also be used to some extent during fasting. In general, the brain can only use water soluble compounds; free fatty acids are not oxidized.

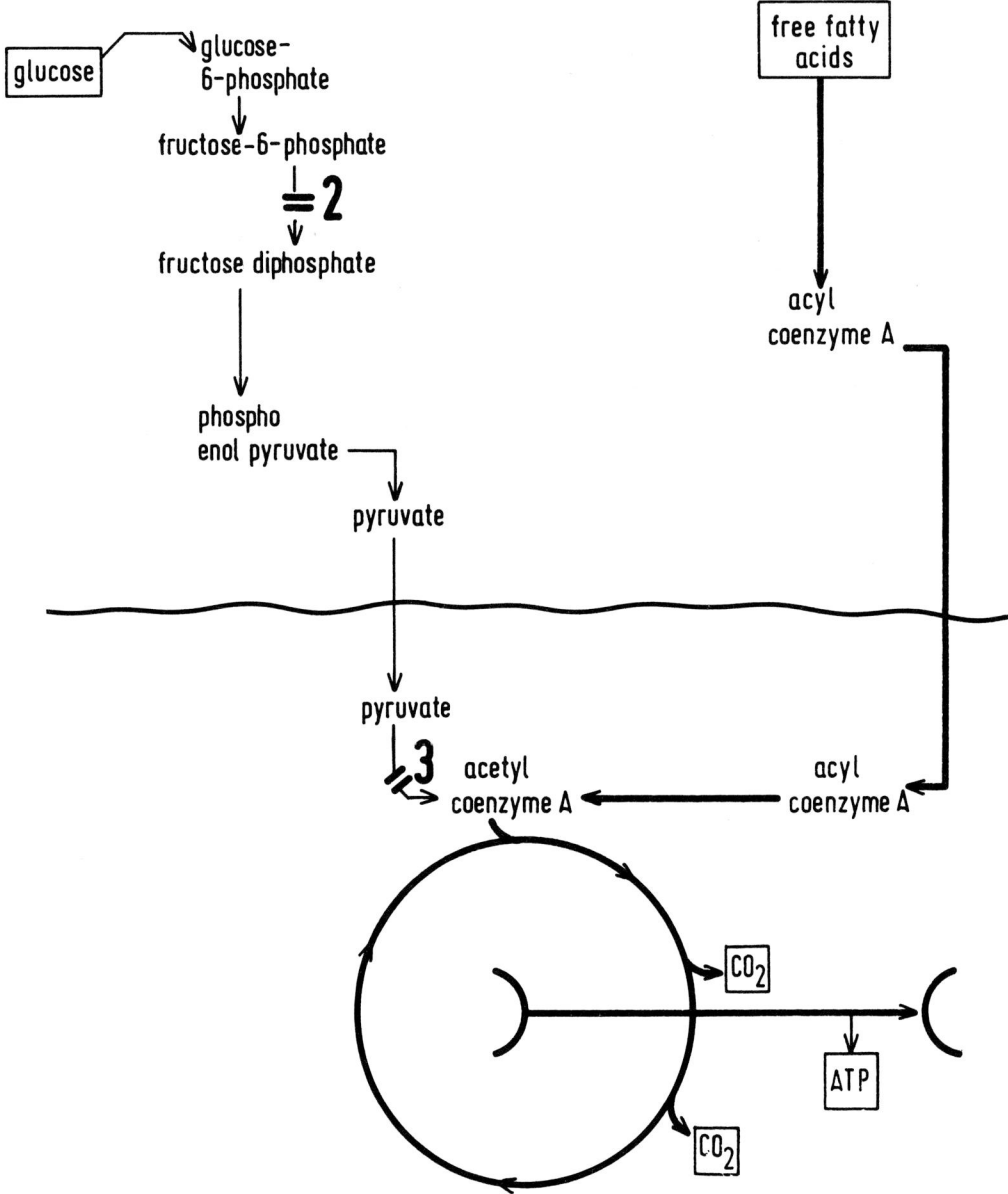

Fig. 6.2 *Oxidation of free fatty acids*

6.2 Oxidation of free fatty acids

The oxidation of free fatty acids is by the pathways shown in Fig. 6.2. The rate is controlled by the availability of free fatty acids, and is therefore determined by those hormones that influence lipolysis, e.g. insulin and adrenaline (section 5.1). The major control during fasting is insulin because its concentration in blood is low and there is no inhibition of the rate of lipolysis in adipose tissue. Thus there is a plentiful supply of free fatty acids to all tissues.

Free fatty acids are oxidized in preference to glucose by skeletal and cardiac muscle because of internal controls. Oxidation of free fatty acids at a high rate produces large quantities of acetyl coenzyme A and citrate. Acetyl coenzyme A inhibits pyruvate dehydrogenase (step 3) and citrate inhibits phosphofructokinase (step 2) so that the rate of glycolysis is reduced.

In liver glycolysis is inhibited by a similar mechanism. In addition the high concentration of acetyl coenzyme A stimulates pyruvate carboxylase (step 5) so that gluconeogenesis from some amino acids and lactate is increased.

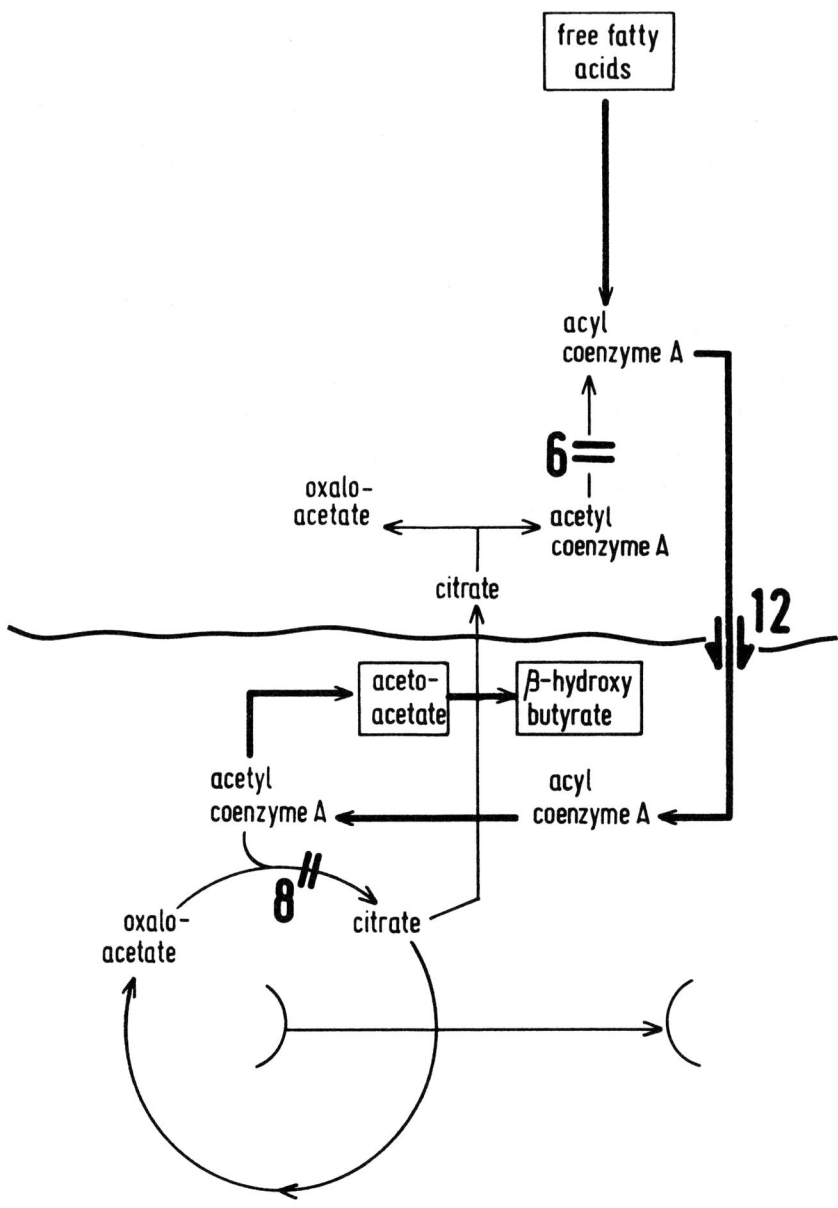

Fig. 6.3 *Formation of ketone bodies*

6.3 Formation of ketone bodies

If the controls on lipolysis cause free fatty acids to be supplied at a rate greater than they can be oxidized (when insulin concentrations are low) liver converts some of them into acetoacetate and β-hydroxybutyrate. The internal controls act by means of the high concentrations of acyl coenzyme A. First the citric acid cycle is limited at step 8 and secondly free fatty acid synthesis is inhibited at step 6. In the liver, only, the low concentration of insulin also stimulates the flow of acyl units into the mitochondria at step 12. Thus some acetyl coenzyme A is channelled into acetoacetate formation (Fig. 6.3).

Acetoacetate and β-hydroxybutyrate are released from the liver and are used by muscle in preference to glucose. During overnight fasting approximately equal amounts of acetoacetate and β-hydroxybutyrate are released, but during long term fasting the total production rate trebles and about 85% of the ketone bodies are β-hydroxybutyrate.

High concentrations of glucagon and adrenaline and a low concentration of insulin stimulate lipolysis and are therefore responsible for high rates of ketone body synthesis.

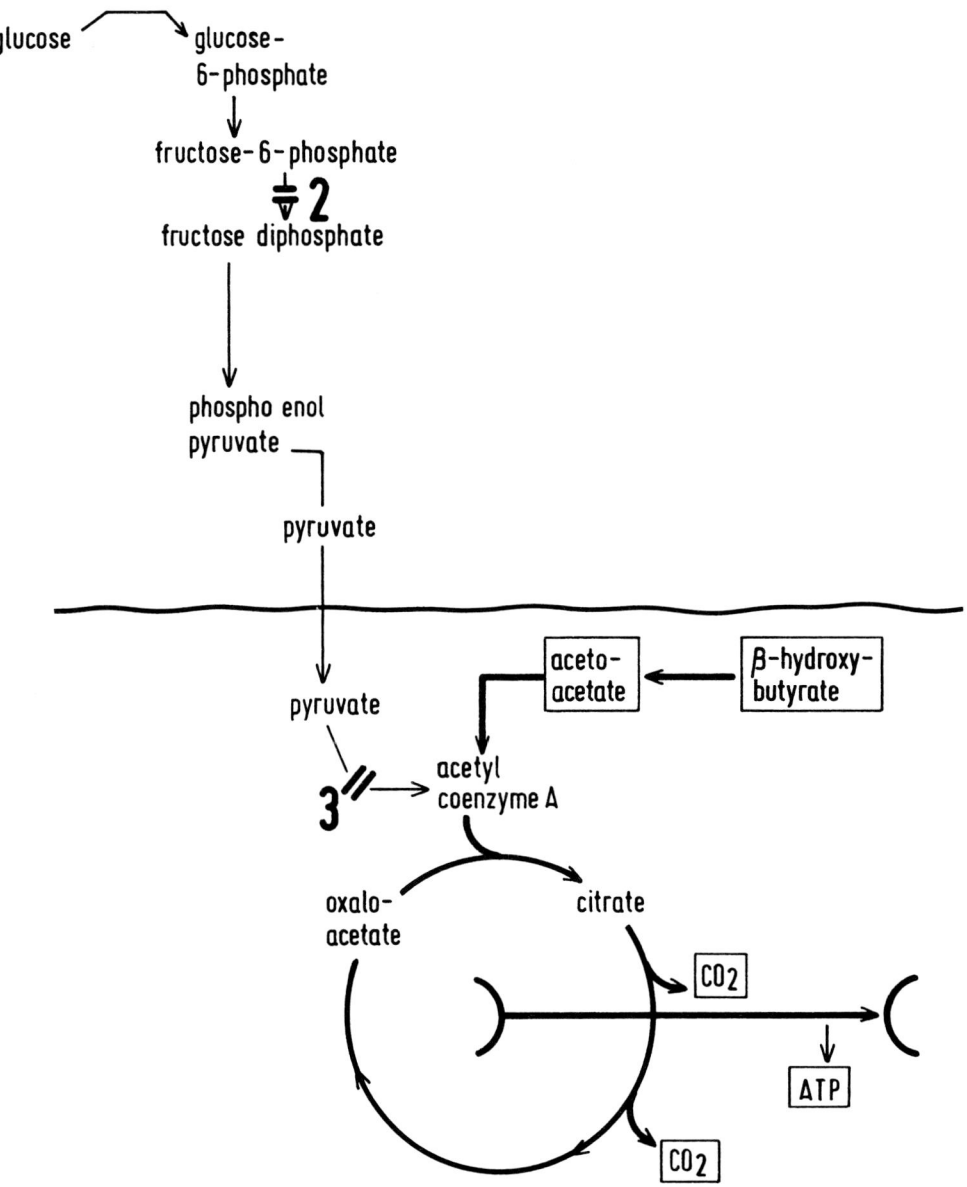

Fig. 6.4 *Oxidation of ketone bodies*

6.4 Oxidation of ketone bodies

β-hydroxybutyrate is converted into acetoacetate which enters the citric acid cycle via acetyl coenzyme A. Thus oxidation is the only metabolic route available to the ketone bodies. Increased concentrations of acetyl coenzyme A and citrate will inhibit glycolysis at steps 2 and 3 (Fig. 6.4). Muscle tissues therefore use free fatty acids and ketone bodies in preference to glucose.

Nervous tissues have the enzymes to use ketone bodies but they are only used to any extent during fasting when the blood concentrations of these fuels increase. Increased availability and oxidation of these materials causes glycolysis, and hence the need for glucose, to diminish.

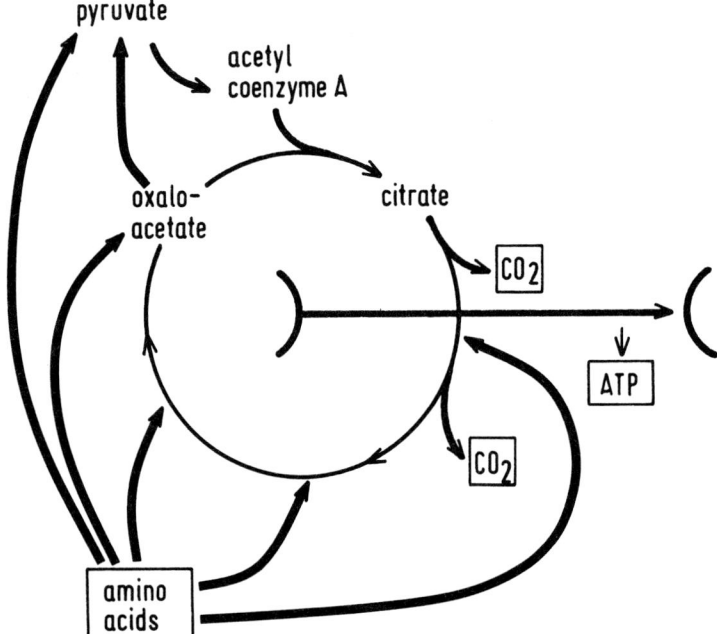

Fig. 6.5 *Oxidation of amino acids*

6.5 Oxidation of amino acids

Provided there is no reduction in the rate of oxidation through the citric acid cycle by a raised concentration of ATP, oxidation of amino acids is by the routes shown in Fig. 6.5. Most of this oxidation takes place in the liver when excess amino acids are available shortly after a meal. Brain, however, during starvation can obtain one third of its energy from amino acid catabolism.

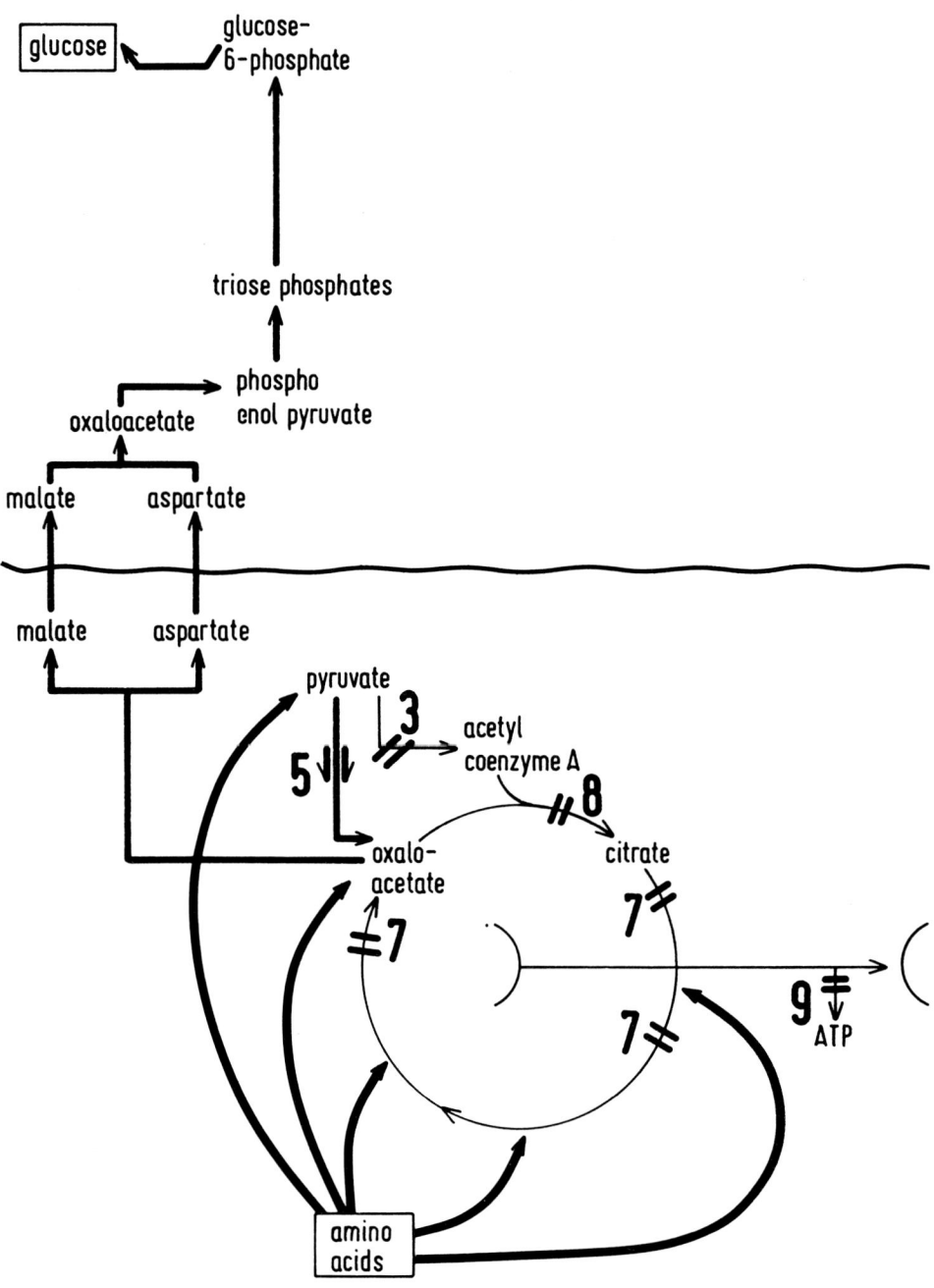

Fig. 6.6 *Gluconeogenesis from amino acids*

6.6 Gluconeogenesis from amino acids

The only tissues which can convert amino acids into glucose are liver and kidney. When this occurs sufficient energy is being produced through the citric acid cycle and oxidative phosphorylation. The rate of oxidation is limited by raised concentrations of ATP (step 9) and NADH (steps 7 and 8). In addition, pyruvate carboxylase is stimulated by an increased concentration of acetyl coenzyme A (step 5). Thus amino acids are directed into gluconeogenesis via oxaloacetate. During the first day or two of a fast the amino acids come from plasma proteins and visceral tissues. After this time, in response to increased concentrations of cortisol, muscle proteins are the most important source. The amino acids are released mainly as alanine (which is used by the liver) and glutamine (which is used by the kidney).

The process requires energy and this is obtained by β-oxidation of free fatty acids.

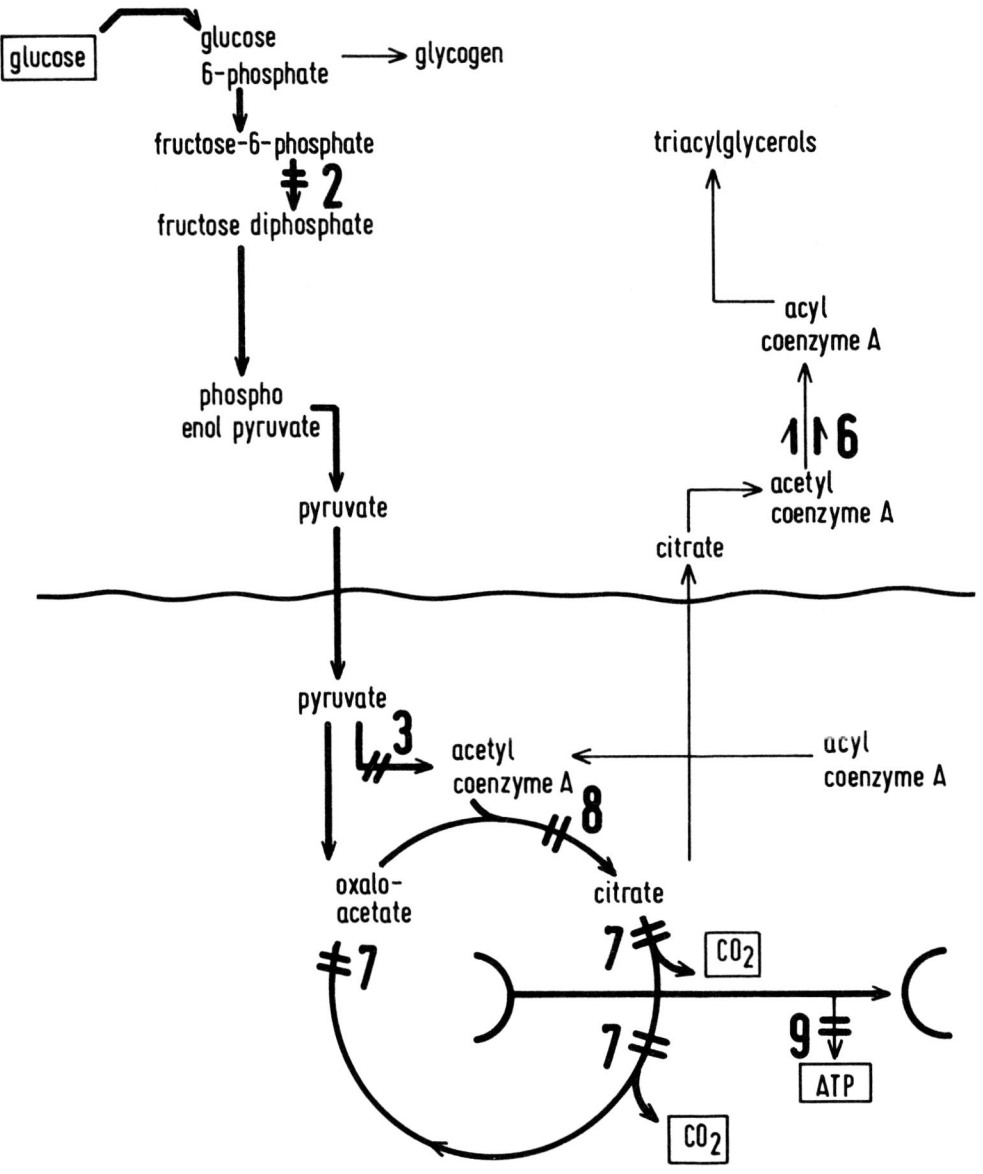

Fig. 6.7 *Oxidation of glucose*

6.7 Oxidation of glucose

Glucose is oxidized through glycolysis, the citric acid cycle and oxidative phosphorylation. The rate is controlled by the demand for energy at step 9 by ATP and at steps 7 and 8 by NADH. In muscle and nervous tissues a high rate of free fatty acid oxidation or ketone body oxidation will also slow glycolysis at steps 2 and 3 (section 6.2). Also there is a control on the rate of uptake of glucose by adipose and muscle tissues. A high blood insulin concentration will cause increased glucose influx. In muscle, due to the controls described above, it will tend to be diverted into glycogen synthesis. In adipose tissue it will be converted into triacylglycerols due to the control at step 6 (section 6.8).

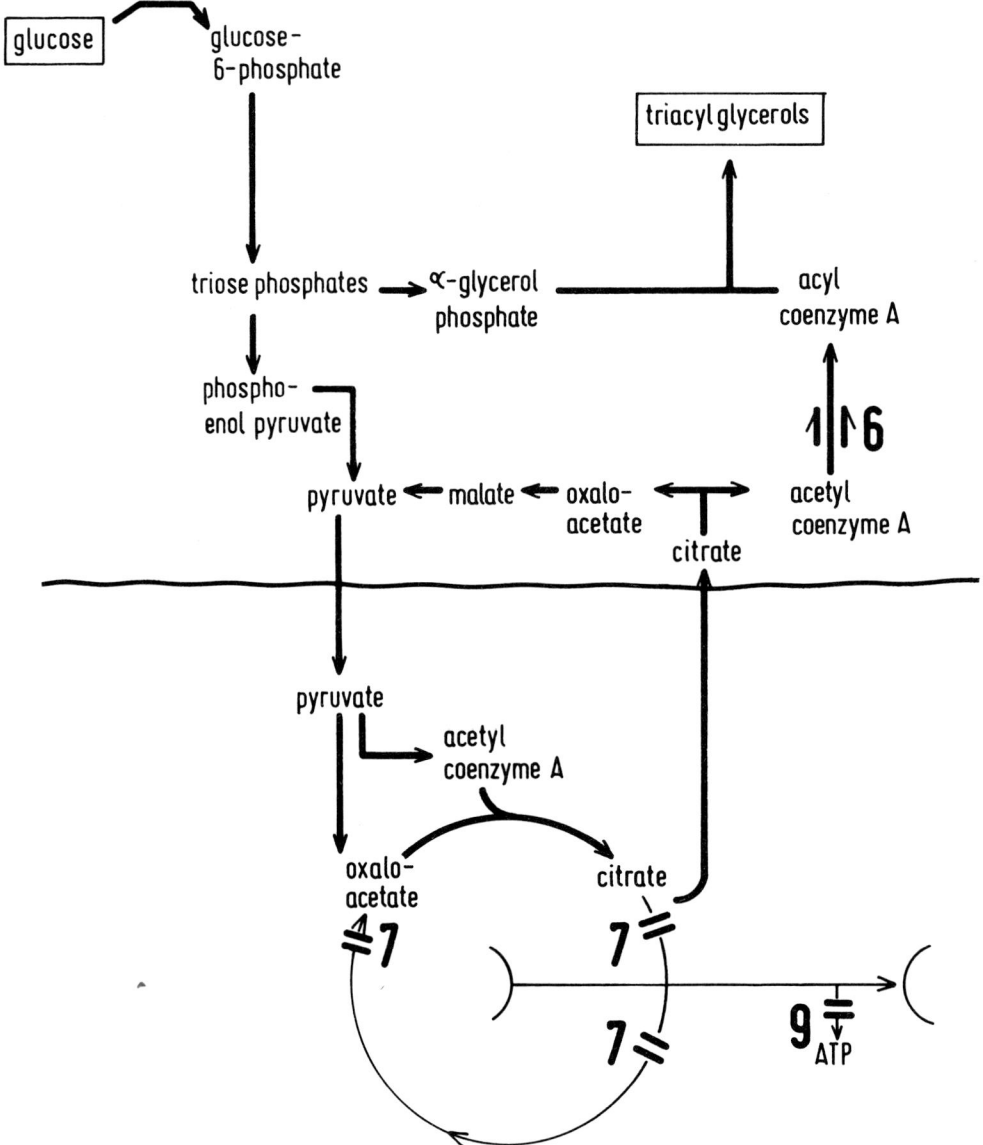

Fig. 6.8 *Conversion of glucose into free fatty acids*

6.8 Conversion of glucose into free fatty acids

When glucose is available in large amounts within liver and adipose tissue, in excess of that needed for energy production, it can be converted into triacylglycerols. In the case of adipose tissue a high concentration of insulin is necessary to cause the increased rate of influx, but with liver the raised glucose concentration itself is sufficient. The rate of oxidation of glucose will be limited by controls within the citric acid cycle and oxidative phosphorylation at steps 9 (ATP) and 7 (NADH). Therefore the concentration of citrate rises and this will stimulate the pathway for free fatty acid synthesis at step 6.

In those tissues which can both synthesize and oxidize free fatty acids, catabolism of incoming free fatty acids inhibits synthesis. The control is by the acyl coenzyme A which inhibit the same key enzyme at step 6.

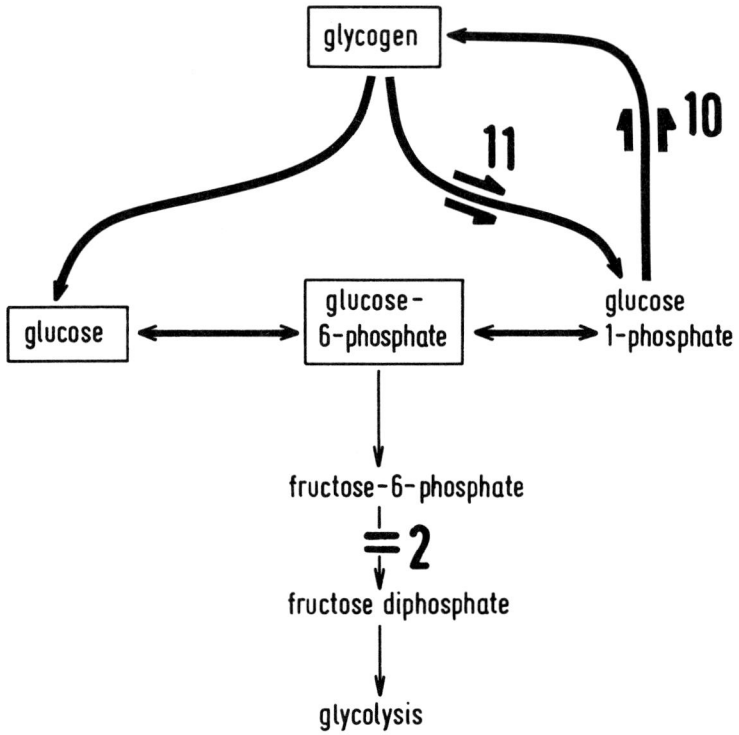

Fig. 6.9 *Glycogen metabolism*

6.9 Glycogen metabolism

Glucose is converted into glycogen by liver and muscle tissues when the influx of glucose is high. With liver a high blood glucose concentration will increase the influx rate, but with muscle a high insulin concentration is also necessary. In both tissues the rate through glycolysis will be limited by the energy demand (section 6.7). Insulin has an additional effect on both tissues by stimulating glycogen synthetase (step 10).

Glycogen catabolism is also controlled mainly by hormones. Adrenaline and glucagon will activate glycogen phosphorylase (step 11). Thus glucose is released by liver to provide a fuel for other tissues. With muscle, glucose-6-phosphate is used internally in glycolysis usually to meet sudden short-term demands.

7. WHOLE BODY METABOLISM

Hormone	Concentration				
	During digestion	Fasting			Exercise
		6–24 hours	2–4 days	2 weeks	
insulin	high	low	low	low	low
cortisol	low	low	high	high	low
adrenaline	low	high	high	high	high
glucagon	low	low	high	low	high

Table 7.1 *Hormones concerned with the control of whole body energy metabolism*

7.1 Whole body metabolism

The various tissues in the body are each specialized in their own functions. The tissues are interdependent and for the whole body to function satisfactorily their functions must be integrated and controlled. Integration and control is by hormones circulating in the blood and their main effect is on the uptake and release of metabolic fuels by the various organs. To a lesser extent, they may have some control on certain key enzymes within the cells.

The release and the effects of hormones have been described in section 5.1. Table 7.1 re-presents the information in a summarized form for use with this section.

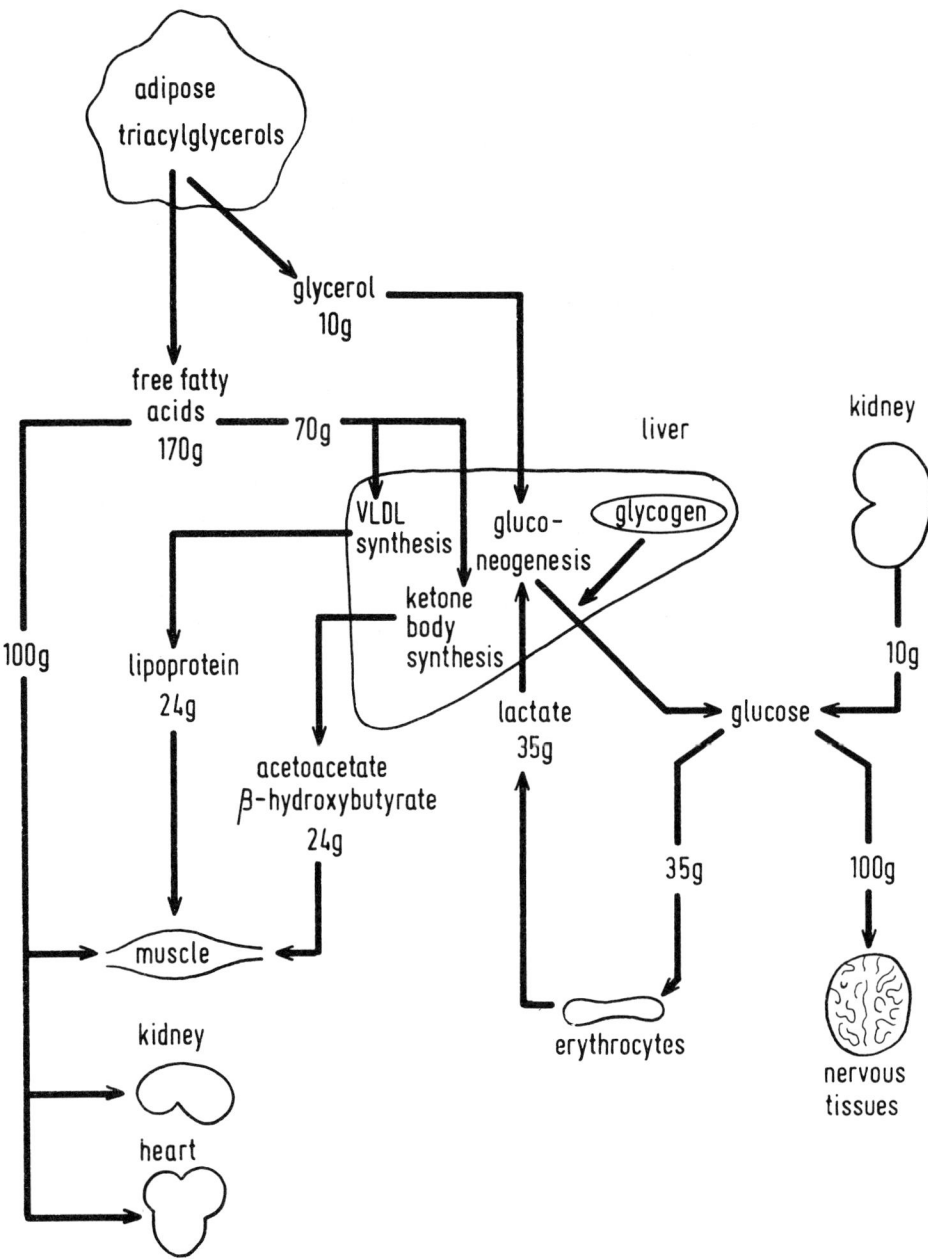

Fig. 7.2 *Flow of metabolic fuels during fasting: 6 to 24 hours*

7.2 The fasted state: 6 to 24 hours

Fig. 7.2 shows the more important sources and uses of the major metabolic fuels during the first 24 hours of a fast. The flow rates are given in g/day. The major energy source is free fatty acids released by adipose tissue. Liver produces most of the glucose and during the first 24 hours it comes predominantly from glycogen breakdown. Only a small quantity is produced by gluconeogenesis, from lactate and glycerol, both by the liver and the kidneys. Small amounts of acetoacetate, β-hydroxybutyrate and lipoproteins are produced by the liver from free fatty acids.

The majority of peripheral tissues use free fatty acids and small amounts of ketone bodies. The lipoproteins are used mainly by skeletal and cardiac muscles. Glucose is used by erythrocytes, which depend entirely upon anaerobic glycolysis, and brain. The latter at this stage of a fast uses approximately 100 g/day.

The major control which maintains this pattern of flow is insulin. The low concentration of insulin allows a high rate for the release of free fatty acids and ensures a low rate for glucose uptake by the peripheral tissues.

In this and the next two figures, the values for the flow rates are very approximate because different sources vary widely in their estimates.

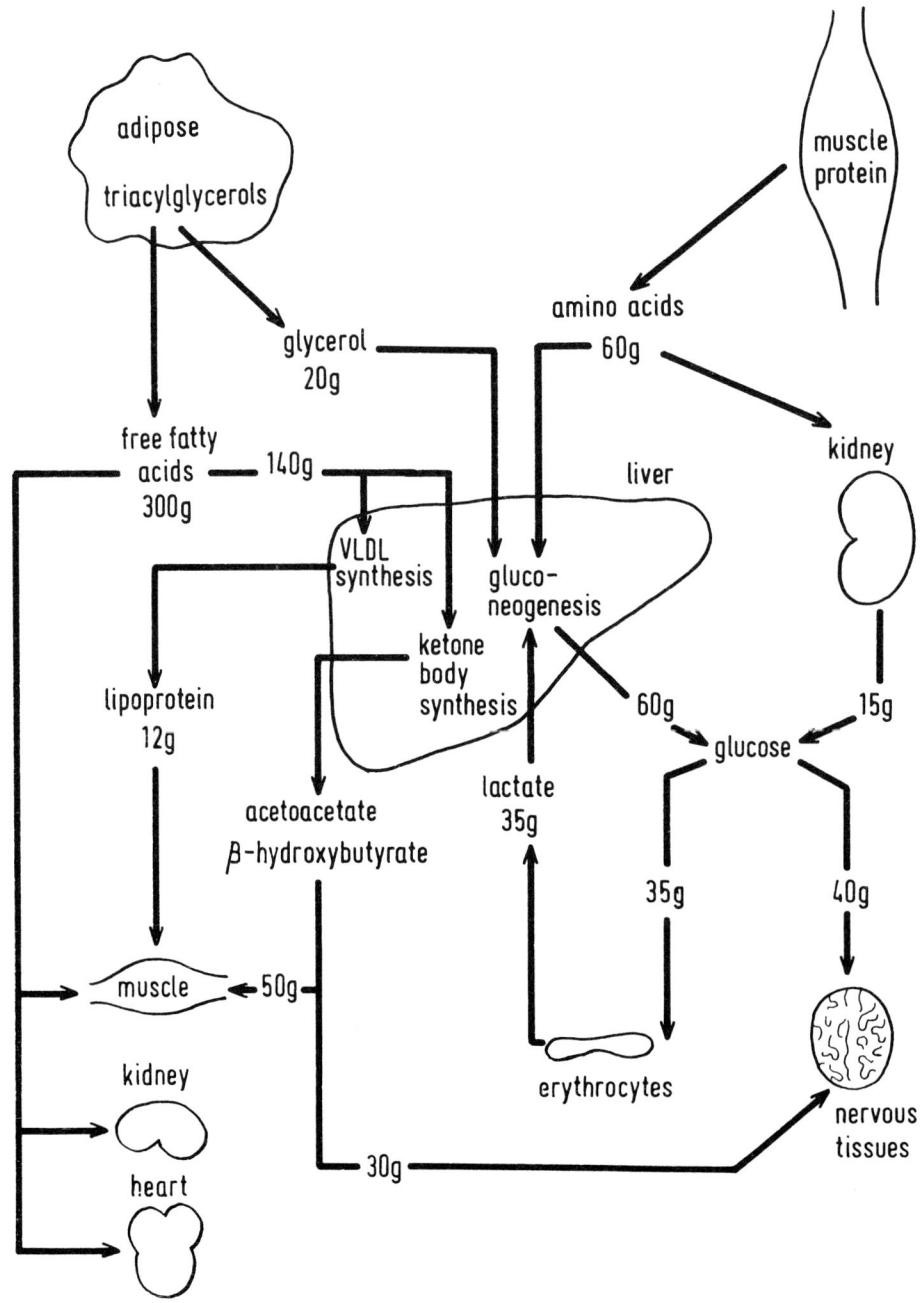

Fig. 7.3 *Fasted: 2 to 4 days*

7.3 The fasted state: 2 to 4 days

By this period of the fast the glycogen store in the liver has been exhausted and glucose is now provided entirely by gluconeogenesis. The precursors for this process are amino acids (mainly from muscle protein), glycerol (from adipose tissue) and lactate (from erythrocytes).

The rate of release of glucose has fallen and the rate of supply of free fatty acids has increased to keep the total energy release approximately constant. Much of the additional free fatty acids is converted by the liver into ketone bodies. A large proportion of these is used by muscle tissues, but brain and nervous tissues are now able to use the remainder in order to replace some of their previous requirement for glucose.

The insulin concentration is still low and its effects are as before. The concentrations of cortisol and adrenaline have risen and they are responsible for the additional release of free fatty acids. The increased concentration of cortisol causes a decreased rate of protein synthesis so that amino acids are now released by several tissues, muscle being quantitatively the most important. Glucagon concentration also increases during this time and at about 4 days it has reached a peak. This hormone causes the rate of gluconeogenesis to increase and also to reach a maximum at about the same time.

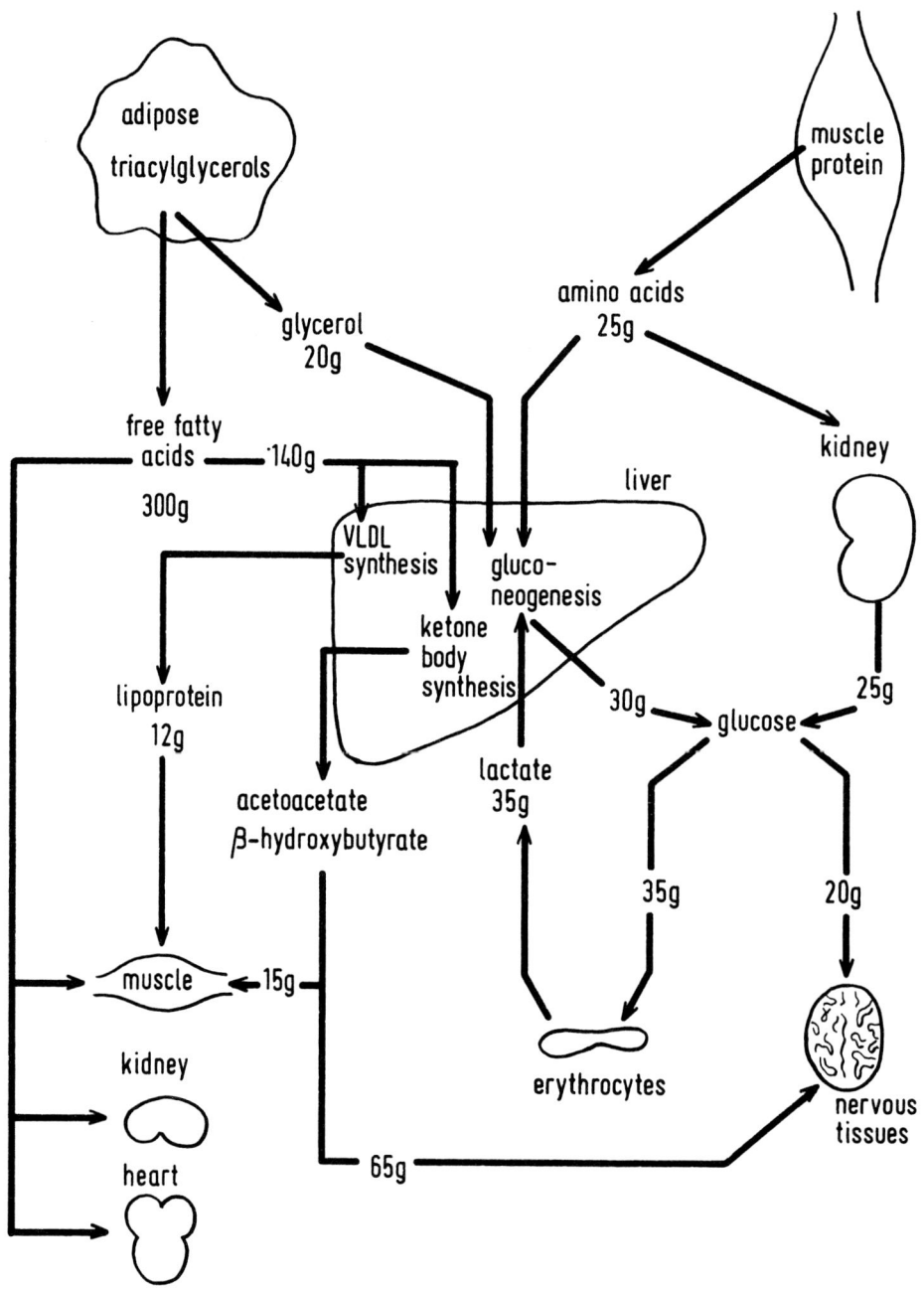

Fig. 7.4 *Fasted: over 2 weeks*

7.4 The fasted state: over 2 weeks

The major change during this period is that the ketone bodies have almost replaced glucose as the fuel for brain and nervous tissues. Ketone body synthesis by the liver is maintained at a high rate. Muscle tissues revert to using mainly free fatty acids so that less ketone bodies are used and the brain is able to use the major portion. Since the requirement for glucose has fallen the rate of gluconeogenesis is reduced and the demand for amino acids falls. Therefore these changes act as a protein sparing mechanism.

During this time the concentration of insulin remains low and those for cortisol and adrenaline remain raised. The major change is that glucagon concentration has fallen and this is largely responsible for the decreased rate of gluconeogenesis.

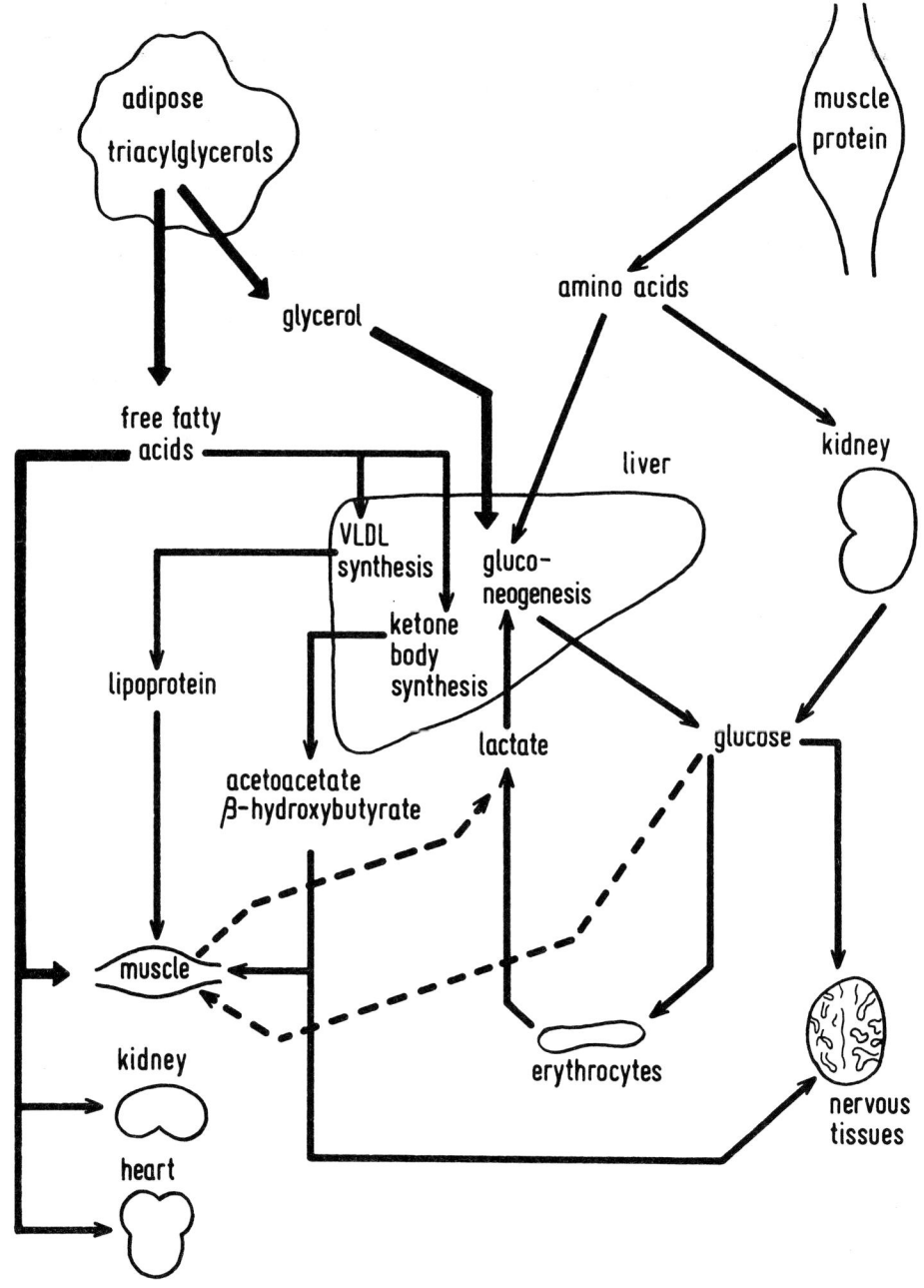

Fig. 7.5 *Flow of metabolic fuels during fasting and exercise*

7.5 Fasted and exercised

During the initial stages (up to 10 minutes) of any sudden demand energy is supplied by the anaerobic conversion of muscle glycogen to lactate. The latter is released to the blood and taken up by the liver. Glucose is resynthesized and recycled.

For about 30 minutes thereafter about half the energy is obtained from blood glucose. In spite of the low insulin concentration, sufficient glucose is able to enter. The remainder of the demand for energy is met mainly by an increased release of free fatty acids. If the large demand for energy continues, free fatty acids increasingly supply the major part of it.

The two important controls are the increasing concentrations of adrenaline and glucagon. Adrenaline causes the release of free fatty acids and the release of glucose from glycogen if available. Glucagon also stimulates the release of glucose.

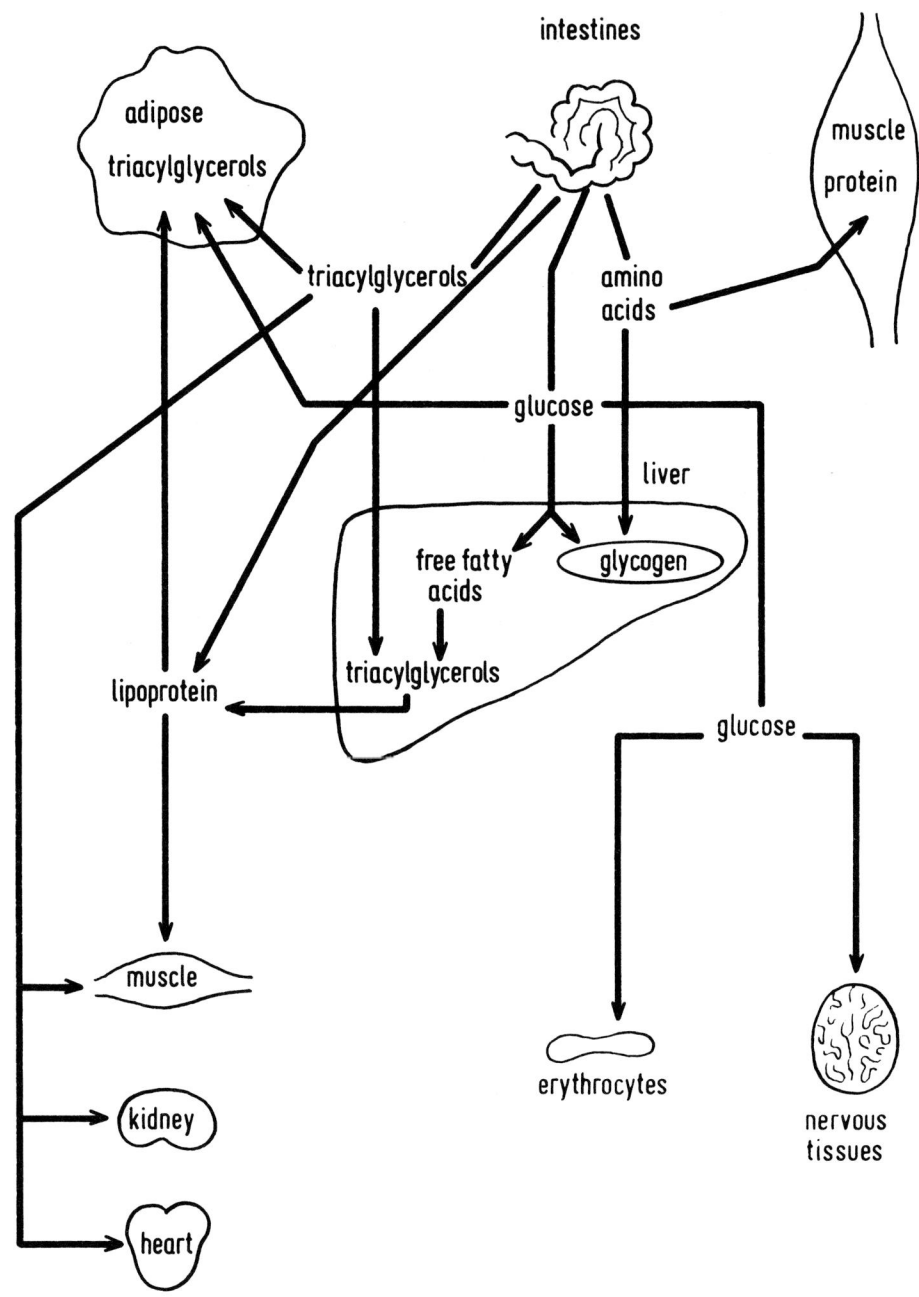

Fig. 7.6 *Flow of metabolic fuels during digestion and absorption*

7.6 The fed state

During digestion the three important fuels are being absorbed from the intestines.

Glucose is absorbed and its high blood concentration causes insulin release from the pancreas. Over half the glucose is taken up by the liver where the insulin causes an increased rate of glycogen synthesis although it has no effect on glucose entry. In adipose tissue (where it is converted into triacylglycerols) and muscle (which stores it as glycogen) the high insulin concentration stimulates the rates of glucose uptake. Only a small proportion is taken up by the other peripheral tissues.

Triacylglycerols, after absorption, are converted by the gut mainly into chylomicra but also partly into lipoproteins. Some of the chylomicra are taken up by the liver where they are partly oxidized and partly converted into more lipoproteins for distribution to other tissues. Adipose tissue will take up most of them and store them as triacylglycerols. The remainder are used by the other peripheral tissues, particularly muscle, and oxidized in place of free fatty acids from storage. The high insulin concentration will activate lipoprotein lipase and accelerate the removal of lipoproteins.

Amino acids are partly used to replace degraded protein, the synthesis being stimulated by the elevated insulin concentration. Some enter gluconeogenesis and the remainder are oxidized.

APPENDIX

Type	Substance	Example of function
amino acids	isoleucine leucine valine lysine methionine threonine phenylalanine tyrosine	protein synthesis
cations	Mg^{2+} Mn^{2+} Zn^{2+} Cu^{2+} Fe^{2+}	hexokinase pyruvate dehydrogenase lactate dehydrogenase amino acid catabolism haemoglobin, cytochromes
B vitamins	thiamine riboflavin nicotinamide pyridoxine folic acid B_{12} pantothenic acid biotin	pyruvate dehydrogenase flavoproteins NAD^+, $NADP^+$ transamination amino acid catabolism amino acid catabolism coenzyme A carboxylations

Table A.1 *Some essential compounds for energy metabolism*

A.1 Some essential compounds for energy metabolism

Besides a sufficient supply of fuels, energy metabolism requires several cofactors for the proper function of many processes. The more important ones are listed in Table A.1.

A correct proportion of essential amino acids in sufficient quantities is necessary for an adequate rate of protein synthesis (section 4.3). The eight listed are those required by an adult – children may need a few additional ones. These amino acids are essential because the body cannot make them from other amino acids by transamination, the main reason being that pathways are not present to synthesize their corresponding keto acids (section 2.7).

In addition to an obvious need for sodium, potassium, calcium, chloride and phosphate, the body also requires certain divalent cations. They form an essential part of the active centre of many enzymes and a few examples are given in Table A.1.

The last group of compounds is the water soluble B vitamins. They are essential because the body cannot synthesize them. Their role in metabolism is that they are converted into the coenzymes necessary for many enzymic reactions. The function of a coenzyme is to act as a carrier for a particular chemical group, e.g. coenzyme A for acyl and acetyl groups, and NAD^+ for hydrogen atoms.

Some suggestions for further reading

Datta, S. P. & Ottaway, J. H. (1976), *Biochemistry*. 3rd ed. Baillière Tindall: London. Ch. 9–12: metabolic pathways, Ch. 17: control mechanisms.

Felig, P. (1976), Recent developments in body fuel metabolism. In: *The Year in Metabolism 1975–1976*, pp. 113–136. Plenum Medical Book Company: New York & London.

Kones, R. J. (1974), *Glucose, insulin, potassium and the heart*. Futura Publishing Company: New York.

Krebs, H. A. & Kornberg, H. L. (1957), A survey of the energy transformations in living matter. *Ergebnisse der Physiologie*, **49**, 212–298.

Lehninger, A. L. (1975), *Biochemistry*. 2nd ed. Worth Publishers, Inc.: New York. Ch. 30: whole body metabolism.

McIlwain, H. (1971), Types of metabolic adaptation in the brain. *Essays in Biochemistry*, **7**, 127–158.

Newsholme, E. A. (1976), Carbohydrate metabolism in vivo: regulation of the blood glucose level. *Clinics in Endocrinology and Metabolism*, **5**, 543–578.

Index

Acetoacetate, 4, 19, 36, 37, 57, 60, 61, 62, 63, 78, 80, 84
Acetoacetyl coenzyme A, 36
Acetyl choline, 8
Acetyl coenzyme A, 11, 50, 52, 70:
   amino acids, oxidation of and, 64
   citric acid cycle and, 18, 19, 23, 63
   fatty acid synthesis and, 32–5 *passim*
   gluconeogenesis and, 66, 67
   glucose, oxidation of, 68
   glycolysis, 22–5 *passim*, 63
   ketone bodies and, 36, 37, 60–3 *passim*
   oxidation of fatty acids and, 20, 21, 58, 59
   production of, 14, 15, 22–26 *passim*, 58, 59
Acyl carnitine transferase, 52
Acyl coenzyme A, 20, 21, 32, 34, 42, 43, 50, 52, 58, 60, 61, 68, 70, 71
Adenosine diphosphate *see* ADP
Adenosine monophosphate *see* AMP
Adenosine triphosphate *see* ATP
Adenylate cyclase, 49
Adipose tissue, 33, 35, 38, 39, 43, 56, 57, 59, 69, 71, 78–82 *passim*, 84, 86, 87
ADP, 7, 15, 16, 17, 51, 52
Adrenaline, 48, 49, 59, 61, 73, 76, 81, 83, 85
Alanine, 31, 57
Amino acids, 4, 14, 18, 23, 30–3 *passim*, 44, 45, 48, 50, 57, 59, 91:
   catabolism of, 25, 26–27, 48, 65
   essential, 90
   fasting, functions during, 81, 82, 83, 84, 86, 87
   gluconeogenesis and, 66, 67
   oxidation of, 64, 65
   uptake of, 56
Ammonia, 26, 27
AMP, 49, 51, 52
Aspartate, 26, 27, 30, 31, 50, 66
ATP, 6, 7, 16, 27, 32, 49, 50, 52, 68, 69, 70, 71:
   amino acid oxidation and, 64, 65
   citric acid cycle and, 18
   conservation of energy by, 7, 15, 17
   fatty acid oxidation and, 20, 21, 58
   gluconeogenesis and, 30, 66, 67
   glycolysis and, 22, 23, 24, 25
   ketone bodies, oxidation of and, 62
   production of, 11, 14, 17, 25, 51
   uses of, 8, 9

B_{12}, Vitamin, 90
Biotin, 90
Bisphosphoglycerate, 17
Blood, 21, 38, 57, 59, 69, 85, 87
Brain, 43, 56, 57, 79, 81, 83

Calcium, 91
Carbohydrates, 3, 4, 33, 45
Carbon dioxide, 2, 3, 15, 18, 19, 22, 24, 27, 30, 32, 50, 58, 62, 64, 68
Carboxylase, 52, 59, 67
Carboxylation, 31, 90
Catabolism, 3, 7, 13–17
Cations, 90, 91
Cells, 3, 31, 51:
   structure of, 10, 11
Chloride, 91
Cholesterol, 39
Cholesterol esters, 39
Chylomicra, 87
Citrate, 6, 11, 18, 19, 30, 32–5 *passim*, 50, 52, 59, 60, 62, 63, 64, 66, 70, 71
Citrate-malate cycle, 33, 35
Citrate synthetase, 52
Citric acid cycle, 7, 11, 14, 17, 18–19, 24, 27, 31, 37, 61, 63, 66, 71:
   energy conservation in, 6
   enzyme control in, 52
   oxidation within, 21, 23, 25, 51, 65, 67, 68
CNS, 56
Coenzyme A, 20, 32, 90, 91
Coenzymes, 7, 11, 17, 19, 91 *see also under names of*
Copper, 90
Cortisol, 48, 53, 67, 76, 81, 83
Creatine, 7
Cristae, 11
Cytochromes, 90
Cytoplasm, 10, 11, 22, 23, 31, 33, 35, 43

Decarboxylation, 31
Dehydrogenases, 51, 52
Digestion, 2, 3, 11, 23

Embden-Meyerhof pathway *see* Glycolysis
Endoplasmic reticulum, 10, 11, 43
Endothelial cell layer, 21
Energy conservation, 6, 7, 15, 17, 19, 23
Enzymes, 11, 31, 37, 43, 44, 57, 63, 91
   control of, 52, 53
   control of metabolism by, 49, 51
   synthesis of, 8, 53
Erythrocytes, 11, 78, 79, 80, 81, 82, 84, 86

Fasting, 23, 31, 37, 39, 45, 53, 57, 59, 61, 65, 67, 76, 78–85
Fat, 3, 5, 43, 57
Fatty acids, 4, 14, 19, 42, 48, 50, 63, 78, 79, 80, 82, 83, 84, 85, 86:
   catabolism of, 25, 71
   enzyme control, 52
   ketone bodies and, 60, 81
   oxidation of, 20, 58, 59, 61, 67, 69, 71
   synthesis of, 7, 11, 32–5, 38, 39, 43, 53, 56, 70, 71
   transport of, 39
   uptake of, 38, 56
Flavoprotein, 6, 90 *see also* fp, fpH$_2$
Folic acid, 90
fp, 6, 16, 17
fpH$_2$, 6, 7, 14, 15, 16, 17, 18, 19, 20, 21
Fructose, 4, 22, 23, 24, 34, 50, 52 *see also following entries*
Fructose-6-phosphate, 22, 24, 30, 34, 50, 58, 62, 68, 72
Fructose bisphosphatase/diphosphatase, 52
Fructose bisphosphate/diphosphate, 22, 23, 30, 32, 50, 58, 62, 68, 72
Fuels, types of, 4, 5
Fumarate, 6, 15, 18, 19, 26

Galactose, 4, 23
Glucagon, 48, 49, 61, 73, 76, 81, 83, 85
Gluconeogenesis, 23, 30, 31, 45, 48, 52, 53, 57, 59, 66, 67, 78–84 *passim*
Glucose, 4, 32, 50, 57, 58, 62, 86, 87:
   fasting, functions during, 78–85 *passim*
   fatty acid synthesis from, 7, 32, 34, 38, 39, 43, 70, 71
   oxidation of, 23, 68, 69
   release of, 48, 85
   resynthesis of, 25, 31, 85
   storage of, 33, 43
   synthesis of, 25, 56, 79, 81
   transport of, 11
   uptake of, 48, 56, 69, 87
   see also following entries, Gluconeogenesis, Glycogen, Glycolysis, Triacylglycerols
Glucose-1-phosphate, 42, 72
Glucose-6-phosphate, 22, 24, 32, 34, 42, 50, 52, 58, 62, 66, 68, 70, 72
Glutamate, 26, 27, 31
Glutamine, 31, 57, 67
Glycerol, 11, 23, 30, 31, 39, 42, 43, 57, 78, 79, 80, 81, 82, 84
α-Glycerol phosphate, 43, 50, 70
Glycogen, 4, 8, 11, 42, 43, 45, 48, 49, 56, 57, 68, 69, 72, 73, 78–82 *passim* 84–7 *passim*
Glycogenolysis, 23, 48, 49
Glycogen phosphorylase, 43, 49, 73
Glycogen synthetase, 43
Glycolysis, 14, 15, 17, 22–5, 27, 35, 42, 43, 52, 59, 63, 68, 72
   anaerobic, 17, 24, 25, 31, 43, 79, 85
Golgi apparatus, 10
Granules, 10

Haemoglobin, 90
Heart, 78, 80, 82, 84, 86 *see also* Muscle tissue, cardiac
Hexokinase, 52, 90
Hexose monophosphate shunt, 35
Hexoses, 14, 15, 19, 23 *see also under names of*
High energy compounds, 6, 7–8
Hormones, 48, 49, 59, 73, 76, 81 *see also under names of*

Hydrogen, 3, 7, 15, 19, 33, 91
Hydrolysis, 3, 5, 17, 21
β-hydroxybutyrate, 4, 36, 37, 57, 60, 61, 62, 63, 78, 79, 80, 82, 84
Hyperglycaemia, 48
Hypoglycaemia, 48, 57

Insulin, 11, 39, 48, 59, 61, 69, 71, 73, 76, 79, 81, 83, 85, 87
Intestines, 3, 23, 43, 86, 87
Ion transport, 9
Iron, 90
Isocitrate, 6, 18
Isoleucine, 90

Keto acids, 91
Ketone bodies, 36, 37, 48, 56, 57, 60–3 *passim*, 69, 78–84 *passim*, see also under names of
Kidneys, 31, 67, 78, 79, 80, 82, 84, 86
Krebs cycle *see* Citric acid cycle

Lactate, 4, 25, 30, 31, 56, 57, 59, 78–82 *passim*, 84, 85
Lactate dehydrogenase, 90
Lactose, 4
Leucine, 90
Lipolysis, 59, 61
Lipoprotein, 3, 21, 38, 39, 56, 78, 79, 80, 82, 84, 86, 87
Lipoprotein lipase, 21, 39, 87
Liver, 36–9 *passim*, 43, 61, 65, 73:
    fasting, functions during, 78–87 *passim*
    fatty acids synthesis and, 33, 35
    functions of, 56, 57
    glucose, synthesis of, 23, 25, 31, 49, 67, 71, 79
Lysine, 90

Macromolecules, 2, 3
Magnesium, 90
Malate, 6, 18, 30, 31, 32, 35, 50, 66

Manganese, 90
Metabolism:
    control mechanisms, 48–53
    essential compounds for, 90, 91
    summary of, 2–11, 56, 57
Metabolites, 51
Methionine, 90
Mitochondria, 10, 11, 17, 19, 20, 22, 23, 24, 31, 33, 35, 52, 61
Muscle tissue, 11, 43, 44, 45, 61, 63, 69, 73, 78, 80, 82, 83, 84, 86, 87:
    cardiac, 39, 56, 57, 59, 79
    skeletal, 31, 39, 56, 57, 59, 79

NAD^+, 6, 16, 17, 36, 90, 91
NADH, 6, 7, 14–25 *passim*, 30, 32, 34, 35, 36, 37, 51, 52, 69
$NADP^+$, 6, 90
NADPH, 6, 7, 33, 34, 35
Nervous system, central, 57
Nervous tissue, 37, 63, 69, 78, 80–4 *passim*, 86
Nicotinamide, 90
Nicotinamide adenine dinucleotide *see* NAD^+, NADH
Nicotinamide adenine dinucleotide phosphate *see* $NADP^+$, NADPH

Oxaloacetate, 6, 11, 14, 15, 18, 26, 30–5 *passim*, 50, 60, 62, 64, 66, 68
Oxidation, 3, 11, 15, 20–1, 37, 51, 57, 58, 59, 61, 62, 63, 64, 65, 68, 69, 87 *see also following two entries, reoxidation and under* Amino acids, Citric acid cycle, Fatty acids, Glucose
β-oxidation, 14, 20–1, 33, 37, 67
Oxidative phosphorylation, 14, 16, 17, 19, 24, 25, 39, 51, 52, 67, 69, 71
α-oxoglutarate, 6, 14, 15, 18, 19, 26, 27, 30, 31

Pancreas, 87
Pantothenic acid, 90
Pentose phosphate pathway, 35
Peripheral tissues, 79, 87

Phenylalanine, 90
Phosphates, 7, 51, 52, 91
Phosphocreatine, 6, 7
Phosphoenolpyruvate, 17, 22, 24, 30, 31, 50, 58, 62, 66, 70
Phosphofructokinase, 52, 59
Phosphoglycerate, 17
Phospholipids, 11, 38, 39
Phosphorylation, 11, 14, 15, 16, 17, 23, 25, 31, 52 *see also* Oxidative phosphorylation, Substrate level phosphorylation
Plasma membrane, 10, 11
Potassium, 91
Protein, 3, 4, 5, 38, 39, 57, 67, 80–4 *passim*, 86
Pyridoxine, 90
Pyruvate, 11, 15, 17, 22–6 *passim*, 30–4 *passim*, 50, 52, 58, 60, 64, 66, 68, 70
Pyruvate carboxylase, 59, 67
Pyruvate dehydrogenase, 59, 90

Reoxidation, 17, 25, 35
Riboflavin, 90

Sodium, 91
Starch, 4
Storage, 2, 3, 7, 33, 41–5, 57
 compounds for, 8
Substrate level phosphorylation, 16, 17, 23, 25
Succinate, 6, 17, 18

Succinyl coenzyme, A, 15, 17, 18, 19, 26
Sucrose, 4
Sympathetic nervous system, 49

Thiamine, 90
Threonine, 90
Tissues, summary of metabolism, 56, 57
Transmitter compounds, 8, 9
Triacylglycerol hydrolysis, 31, 48
Triacylglycerols, 4, 33, 38, 39, 42, 43, 45, 50, 56, 57, 68, 69, 70, 71, 86, 87
Triose phosphates, 22, 23, 24, 30, 42, 50, 66, 70
Tyrosine, 90

Urea and urea cycle, 2, 3, 26, 27, 53

Valine, 90
Visceral organs, 45, 67
Vitamins, 90, 91
VLDL, 38, 39, 78, 80, 82, 84

Water, 3, 15, 17

Zinc, 90